AF677964

Functional Coatings for Metallic and Ceramic Materials

Guest Editors

Michael Zinigrad
Department of Chemical Engineering, Biotechnology and Materials
Ariel University
Ariel
Israel

Alexey Kossenko
Department of Chemical Engineering, Biotechnology and Materials
Ariel University
Ariel
Israel

Editorial Office
MDPI AG
Grosspeteranlage 5
4052 Basel, Switzerland

This is a reprint of the Special Issue, published open access by the journal *Coatings* (ISSN 2079-6412), freely accessible at: https://www.mdpi.com/journal/coatings/special_issues/metallic_ceramic_materials.

For citation purposes, cite each article independently as indicated on the article page online and as indicated below:

Lastname, A.A.; Lastname, B.B. Article Title. *Journal Name* **Year**, *Volume Number*, Page Range.

ISBN 978-3-7258-5227-7 (Hbk)
ISBN 978-3-7258-5228-4 (PDF)
https://doi.org/10.3390/books978-3-7258-5228-4

© 2025 by the authors. Articles in this book are Open Access and distributed under the Creative Commons Attribution (CC BY) license. The book as a whole is distributed by MDPI under the terms and conditions of the Creative Commons Attribution-NonCommercial-NoDerivs (CC BY-NC-ND) license (https://creativecommons.org/licenses/by-nc-nd/4.0/).

Functional Coatings for Metallic and Ceramic Materials

Guest Editors

Michael Zinigrad
Alexey Kossenko

MDPI

Basel • Beijing • Wuhan • Barcelona • Belgrade • Novi Sad • Cluj • Manchester

Contents

coatings

MDPI

Article

Surface Heterostructure of Aluminum with Carbon Nanotubes Obtained by Laser-Oriented Deposition

Natalia Kamanina [1,2,3,*], Konstantin Borodianskiy [4] and Dmitry Kvashnin [5,6,7,*]

1 Lab for Photophysics of Media with Nanoobjects at Vavilov State Optical Institute, Kadetskaya Liniya V.O., dom 5/2, 199053 St. Petersburg, Russia
2 St. Petersburg Electrotechnical University ("LETI"), Ul.Prof.Popova, dom 5, 197376 St. Petersburg, Russia
3 Scientific Center "Kurchatov Institute"—Nuclear Physics Institute, Orlova Roscha 1, Leningradskij Region, 188300 Gatchina, Russia
4 Department of Chemical Engineering, Ariel University, Ariel 40700, Israel; konstantinb@ariel.ac.il
5 Emanuel Institute of Biochemical Physics RAS, 4 Kosigina st., 119334 Moscow, Russia
6 Pirogov Russian National Research Medical University, Ostrovitianov str. 1, 117997 Moscow, Russia
7 School of Chemistry and Technology of Polymer Materials, Plekhanov Russian University of Economics, Stremyanny Lane 36, 117997 Moscow, Russia
* Correspondence: nvkamanina@mail.ru (N.K.); dgkvashnin@phystech.edu (D.K.); Tel.: +7-(812)-327-00-95 (N.K.)

Abstract: Al is one of the most widely applicable metallic materials due to its advanced properties. However, its main drawback is its strength, which is relatively low compared to ferrous alloys. This issue may be resolved using different approaches. In the present work, a heterostructure of Al substrate with a modified surface with carbon nanotubes (CNTs) was studied. This heterostructure was obtained using the laser-oriented deposition technique. The obtained results showed a slight reduction in the reflectivity of the obtained Al substrate with embedded CNTs compared to pure Al. Additionally, the obtained surface heterostructure showed enhancement in microhardness and higher hydrophobicity. Simulation of the CNT embedding process revealed that CNT penetration strongly depends on the diameter. Hence, the penetration increases when the diameter decreases.

Keywords: aluminum; carbon nanotubes; laser-oriented deposition technique; molecular dynamics

1. Introduction

It is well known that various areas of human activity include aluminum as a primary or secondary material for different applications [1–6]. Aluminum and its alloys are widely used in the automotive and aircraft industry, railway station construction, and shipbuilding, as well as in general optoelectronics and in cryogenic technology. Despite the many advantages of aluminum, the low strength significantly limits its potential applications. Many attempts have now been made to improve the mechanical properties of aluminum by creating nanostructured composites with the addition of two-dimensional materials [7–14]. Thus, the improvement of the mechanical properties of these metallic materials is an issue of high scientific interest.

The basic principles of molecular physics show that the interatomic distance in crystal structures is proportional to the lattice parameter. Aluminum has a face-centered cubic (FCC) lattice with a lattice parameter equal to 4.046×10^{-8} cm at room temperature [15]. Therefore, it is logical to assume that the formation of strong C–C bonds, which exhibit lengths of ~0.13–0.14 nm, and high Young's modulus [16,17] on the surface of Al are beneficial for the advanced performance of the surface.

CO_2 lasers are used for a wide variety of applications. Hence, Liu et al. used this approach to synthesize ceramic material by sintering of alumina with boron carbide additions [18]. Chen showed a sintering process of yttria-stabilized zirconia with MgO powder using a CO_2 laser [19]. Laser annealing was investigated by Jo et al. The authors used a CO_2 laser in the annealing of Al-doped ZnO material to reduce defects that appear during fabrication [20]. The surface heterostructure fabrication using the laser-oriented deposition technique was previously described in [21,22]. The laser-oriented deposition technique was used to improve basic parameters of the inorganic crystals of KBr and LiF with carbon nanotubes (CNTs) and showed an advanced performance of the produced surface heterostructure [23,24].

Here, the innovative laser-oriented deposition technique was applied to produce a heterostructure of CNTs embedded in aluminum substrate. The obtained heterostructure was examined for reflectivity, microhardness, and wetting, and the obtained results were compared with pure Al. Additionally, a molecular dynamic simulation was performed to illustrate the process of surface heterostructure formation.

2. Materials and Methods

The modification of the aluminum surface was performed with an IR CO_2 laser operated at a wavelength of 10.6 μm with a power of 30 W and a beam diameter of 5 mm. The system was connected to a vacuum chamber. CNTs were embedded in the material's surface at an additional electric field of 100–600 $V{\cdot}cm^{-1}$ which was applied to orient the CNTs in the vertical position during the deposition process. The grid to which the voltage was applied was placed at a variable distance in front of the samples. This made it possible to change the electric field's strength and conduct directional oriented CNT deposition. It should be mentioned that this method does not require the creation of any additional conditions for heating of the substrates and the composition of the gas reagents, which favorably differs from the classical methods based on the chemical vapor deposition and physical vapor deposition processes. Thus, the laser-oriented deposition (LOD) method was realized efficiently.

The Al substrate with a diameter of 35 mm and a thickness of 5 mm used in the work is shown in Figure 1. The CNTs used in the work were single-wall carbon nanotubes (SWCNTs) type #704121, with a diameter in the range of 0.7–1.1 nm (Aldrich Co., Karlsruhe, Germany). The spectra of the nano-object-treated Al materials were obtained using the Perkin-Elmer Lambda 9 and the Furrier FSM-1202 instruments ("Nica-Garant+", Saint-Petersburg, Russia). The contact angle (CA) was tested using the OCA 15EC device (LabTech Co., Saint-Petersburg-Moscow, Russia). Contact angle values presented in this work are the average of five points on each of five tested samples.

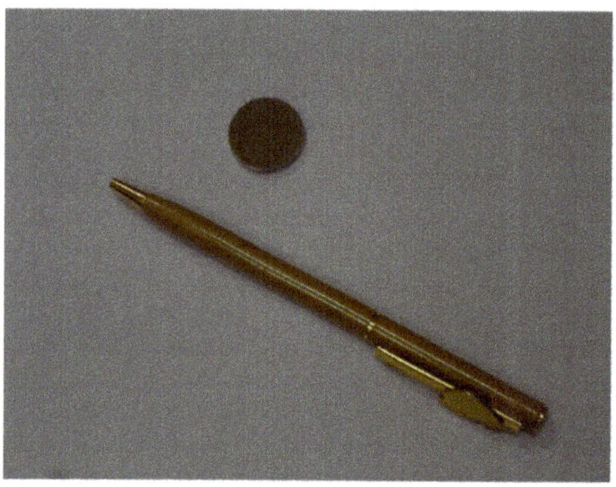

Figure 1. The examined Al substrate.

Simulations of CNT deposition into the Al surface were performed using classical molecular dynamics using the LAMMPS simulation package [25]. The Al(111) surface was considered as the most favorable in terms of energy. Behavior of the Al substrate at a finite temperature was examined by means of the embedded atom method (EAM) [26] which is applicable for metal simulations.

The Al substrate–CNT interaction was described by Lennard-Jones potential according to the equation:

$$E = 4\varepsilon\left[\left(\frac{\sigma}{r}\right)^{12} - \left(\frac{\sigma}{r}\right)^{6}\right] \quad (1)$$

where ε = 2.63 eV, σ = 1.91 Å. CNT behavior at the final temperature was described with a well-developed theoretical approach of the Tersoff many-body potential [27], which was successfully applied in the description of the mechanical properties of graphene membranes under defects [28]. The Al substrate included ~45,000 atoms with a total height of about 9 nm which was enough to protect CNTs from the penetration through the substrate. The molecular dynamics simulation was carried out at a constant temperature of 300 K. The total time of simulation was 60 ps with a time step of 1 fs. The acceleration rate of the CNTs varied from 100 to 600 m/s according to the experimental set up.

3. Results

The reflectance spectra of the pure Al and CNTs embedded in Al substrate are shown in Figure 2. Results were obtained by specular reflection at 45°.

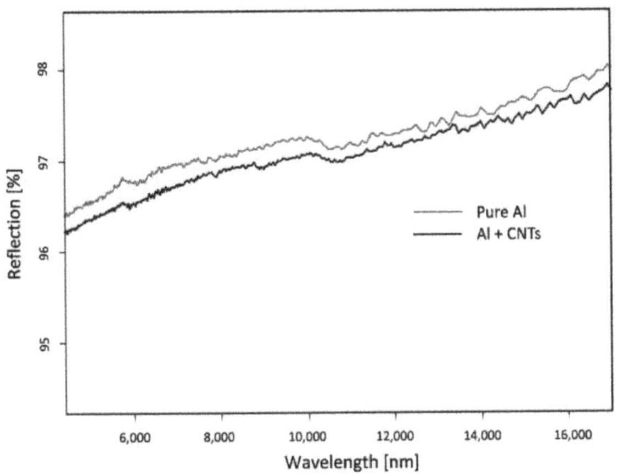

Figure 2. The reflectance spectra of CNTs embedded in Al substrate and pure Al.

The reflectance values at a wavelength range of 6000–16,000 nm of CNTs embedded in Al substrate showed lower values in comparison with the pure Al. This behavior may be attributed to the decrease in the Fresnel losses in the structured system. The obtained data fully correlate with the performance of a smooth aluminum surface whose reflectivity is more than 90% at a wavelength range of 900–12,000 nm, and for a wavelength of 200 nm, the reflectivity was reduced to 70%.

The wettability of the heterostructure surface was examined and compared with the surface of pure Al. The obtained data for CA are shown in Table 1. It was found that the CA of the CNTs embedded in the Al substrate is larger than that of pure Al. This behavior may be attributed to the lotus effect; it is well known that nanostructured materials usually exhibit higher hydrophobicity.

Table 1. Contact angle (CA) data for CNTs embedded in Al substrate and pure Al.

Material	Contact Angle (°)	Increase (%)
Pure Al	108	+3
Al + CNTs	111	

The microhardness values obtained are shown in Table 2. CNTs embedded in the Al substrate had microhardness of 96.3 MPa while the pure Al had a value of 90.3 MPa. This behavior may be attributed to the presence of harder CNTs on the surface of the Al.

Table 2. Microhardness data for CNTs embedded in Al substrate and pure Al.

Material	Microhardness (MPa)	Increase (%)
Pure Al	90.3	+7
Al + CNTs	96.3	

The large-scale molecular dynamic simulation process was performed to understand the atomic structure behavior of CNTs embedded in the Al substrate. The simulation was performed based on CNTs with various diameters in the range of 0.64–3.35 nm which were accelerated from 100 to 600 m/s.

Figure 3 shows the plot of the penetration depth as a function of the CNT acceleration rate. It is clearly seen that the embedding of narrower CNTs led to the more local deformation of the Al surface together with their deeper penetration. Meanwhile, the atomic structure of narrower CNTs is much less sensitive to the structural distortions than wider CNTs. Here, the penetration limit of the wider CNTs is slightly over 2 nm while the penetration of the narrower CNTs reaches 5 nm. This behavior may be attributed to the decrease in curvature of the wider CNTs with a subsequent decrease in the bending modulus. Moreover, the simulation also revealed that the penetration depth of CNTs independent of its orientation towards the substrate surface. The penetration of two nanotubes of a diameter of 0.64 and 0.94 nm was also simulated at 20° related to the surface. Negligible changes in penetration depth are found for both CNTs.

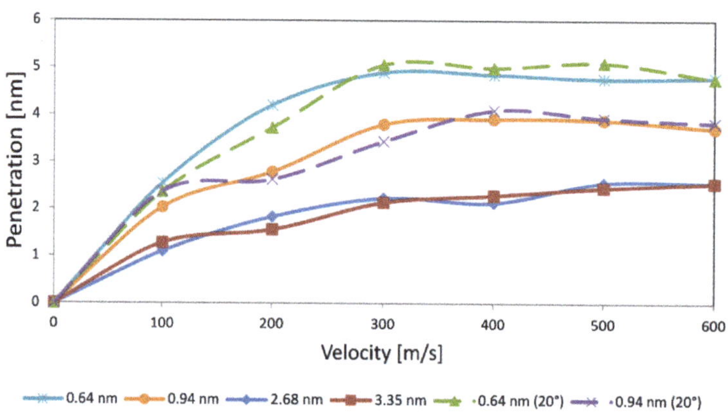

Figure 3. The plot of penetration depth of CNTs in Al substrate as a function of their acceleration rate.

Figure 4a illustrates a schematic presentation of a CNT with a diameter of 0.64 nm embedded in Al substrate and Figure 4b illustrates a schematic presentation of a CNT with a diameter of 2.68 nm embedded in Al substrate. The simulation was performed using an acceleration rate of the CNTs of 300 m/s. As expected, these images revealed that the narrower CNT penetrates deeper into the Al substrate than the wider one.

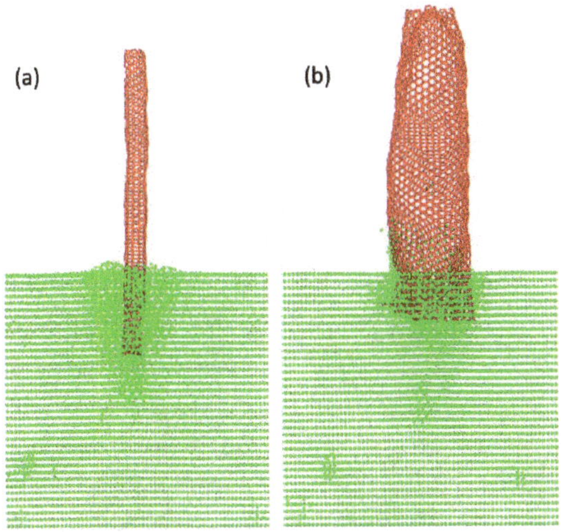

Figure 4. Schematic presentation of the atomic simulated structure of (**a**) CNT with a diameter of 0.64 nm embedded in Al substrate, and (**b**) CNT with a diameter of 2.68 nm embedded in Al substrate. Simulation was done using acceleration rate of 300 m/s.

4. Conclusions

In this work, an Al surface heterostructure was produced by the embedding of CNTs in Al substrate using the laser-oriented deposition technique to improve its mechanical and optical properties. It was shown that the reflectivity of the surface heterostructure exhibits lower values compared to pure Al. The contact angle of the surface of CNTs embedded in Al substrate increased from 108° to 111°, which is well correlated with the hydrophobicity of nanostructures according to the lotus effect. Microhardness tests also revealed a slight improvement in the surface heterostructure by 7%. Finally, a simulation of the CNT embedding process showed that narrower CNTs penetrate deeper into the substrate. Hence, the penetration of wider CNTs, with a diameter of 2.68–3.35 nm, is slightly over 2 nm while the narrower CNTs, with a diameter of 0.64 nm, penetrate over 5 nm.

It should be mentioned that some limit in size to form the coinciding conditions between the metal and CNT is related to the lattice parameter of the substrate and the diameter of the CNT. The diameter of the CNT should be less than the lattice parameter by no more than 0.1–0.2 nm. However, our previous study showed that due to the defects presence in the substrate, which usually lead to lattice deformations, such a limitation cannot be achieved. Therefore, practically, there are no restrictions on the CNT diameter.

Indeed, it should be remarked that other alternative experiments should be carried out to support the covalent bonding between the CNTs and matrix material surface atoms. This will be performed in the future and will be presented in comparison with other materials treated with the LOD technique.

Author Contributions: Conceptualization, N.K.; Formal analysis N.K.; Investigation, N.K.; Methodology, D.K.; Resources, D.K.; Simulation, D.K.; Visualization, K.B.; Writing—original draft preparation, N.K. and K.B. All authors have read and agreed to the published version of the manuscript.

Funding: The experimental results were partially obtained and analyzed via Russian National Base Project "Nanocoatings" (2012–2015). D.G.K. acknowledges the Ministry of Science and Higher Education of the Russian Federation (project No. 01201253304) for partial financial support.

Institutional Review Board Statement: Not applicable.

Informed Consent Statement: Not applicable.

Data Availability Statement: Publicly available data on laser-oriented CNTs deposition can be found in the patent: Kamanina, N.V.; Vasilyev, P.Y.; Studeonov, V.I. Optical Coating Based on Oriented in the Electric Field CNTs for the Optical Devises, Micro- and Nanoelectronics under the Conditions When the Interface: Solid Substrate-Coating Can Be Eliminated. RU Patent 2 405 177 C2, 23.12.2008.

Acknowledgments: The authors would like to thank their colleagues from the labs and institutes for the kind discussion and remarks. In particular, the authors would like to acknowledge P.Ya.Vasilyev for his help in running experiments with the laser-oriented deposition process.

Conflicts of Interest: The authors declare no conflict of interest.

References

1. Johnson, P.B.; Christy, R.W. Optical constants of the noble metals. *Phys. Rev. B* **1972**, *6*, 4370. [CrossRef]
2. Valkonen, E.; Karlsson, B.; Ribbing, C.-G. Solar optical properties of thin films of Cu, Ag, Au, Cr, Fe, Co, Ni and Al. *Solar Energ* **1984**, *32*, 211–222. [CrossRef]
3. Lynch, D.W.; Hunter, W.R. Comments on the optical constants of metals and an introduction to the data for several metals. In *Handbook of Optical Constants of Solids*; Academic Press: New York, NY, USA, 1997; Volume 1, pp. 275–367. [CrossRef]
4. Rakic, A.D.; Djurisic, A.B.; Elazar, J.M.; Majewski, M.L. Optical properties of metallic films for vertical-cavity optoelectroni devices. *Appl. Opt.* **1998**, *37*, 5271–5283. [CrossRef]
5. Scaffardi, L.B.; Lester, M.; Skigin, D.; Tocho, J.O. Optical extinction spectroscopy used to characterize metallic nanowires *Nanotechnology* **2007**, *18*, 315402. [CrossRef]
6. Temple, T.L. Choosing optical constants for the simulation of Ag, Au and Cu nanospheres. *Plasmonics* **2015**, *10*, 455–467. [CrossRef
7. Krasheninnikov, A.V.; Berseneva, N.; Kvashnin, D.G.; Enkovaara, J.; Björkman, T.; Sorokin, P.; Shtansky, D.; Nieminen, R.M Golberg, D. Toward stronger Al–BN nanotube composite materials: Insights into bonding at the Al/BN interface from first principles calculations. *J. Phys. Chem. C* **2014**, *118*, 26894–26901. [CrossRef]
8. Kvashnin, D.G.; Ghorbani-Asl, M.; Shtansky, D.V.; Golberg, D.; Krasheninnikov, A.V.; Sorokin, P.B. Mechanical properties and current-carrying capacity of Al reinforced with graphene/BN nanoribbons: A computational study. *Nanoscale* **2016**, *8* 20080–20089. [CrossRef] [PubMed]
9. Steinman, A.E.; Corthay, S.; Firestein, K.L.; Kvashnin, D.G.; Kovalskii, A.M.; Matveev, A.T.; Sorokin, P.B.; Golberg, D.V.; Shtansky D.V. Al-based composites reinforced with AlB2, AlN and BN phases: Experimental and theoretical studies. *Mater. Des.* **2018**, *141* 88–98. [CrossRef]
10. Konopatsky, A.S.; Kvashnin, D.G.; Corthay, S.; Boyarintsev, I.; Firestein, K.L.; Orekhov, A.; Arkharova, N.; Golberg, D.V.; Shtansky D.V. Microstructure evolution during AlSi10Mg molten alloy/BN microflake interactions in metal matrix composites obtained through 3D printing. *J. Alloys Compd.* **2021**, *859*, 157765. [CrossRef]
11. Borodianskiy, K.; Zinigrad, M. *Nanomaterials Applications in Modern Metallurgical Processes. Diffusion Foundations*; Trans Tech Publications Inc.: Durnten-Zurich, Switzerland, 2016; Volume 9, pp. 30–41. [CrossRef]
12. Borodianskiy, K.; Selivorstov, V.; Dotsenko, Y.; Zinigrad, M. Effect of additions of ceramic nanoparticles and gas-dynamic treatment on Al casting alloys. *Metals* **2015**, *5*, 2277–2288. [CrossRef]
13. Borodianskiy, K.; Zinigrad, M. Mechanical properties and microstructure characterization of Al-Si cast alloys formation using carbide nanoparticles. *J. Mater. Sci. Appl.* **2015**, *1*, 85–90.
14. Zinigrad, M.; Mazurovsky, V.; Borodianskiy, K. Physico-chemical and mathematical modeling of phase interaction taking place during fusion welding processes. *Mater. Sci. Eng. Technol.* **2005**, *36*, 489–496. [CrossRef]
15. Davey, W.P. Precision measurements of the lattice constants of twelve common metals. *Phys. Rev.* **1925**, *25*, 753–761. [CrossRef]
16. Robertson, J. Realistic applications of CNTs. *Mater. Today* **2004**, *7*, 46–52. [CrossRef]
17. Namilae, S.; Chandra, N.; Shet, C. Mechanical behavior of functionalized nanotubes. *Chem. Phys. Lett.* **2004**, *387*, 247–252 [CrossRef]
18. Liu, R.Z.; Chen, P.; Wu, J.M.; Chen, S.; Chen, A.N.; Chen, J.Y.; Liu, S.S.; Shi, Y.S.; Li, C.H. Effects of B_4C addition on the microstructure and properties of porous alumina ceramics fabricated by direct selective laser sintering. *Ceram. Int.* **2018**, *44* 19678–19685. [CrossRef]
19. Chen, F.; Wu, J.M.; Wu, H.Q.; Chen, Y.; Li, C.H.; Shi, Y.S. Microstructure and mechanical properties of 3Y-TZP dental ceramics fabricated by selective laser sintering combined with cold isostatic pressing. *Int. J. Lightweight Mater. Manuf.* **2018**, *1*, 239–245 [CrossRef]
20. Jo, G.H.; Ji, J.H.; Masao, K.; Ha, J.G.; Lee, S.K.; Koh, J.H. CO_2 laser annealing effects for Al-doped ZnO multilayered films. *Ceram Int.* **2018**, *44*, S211–S215. [CrossRef]
21. Kamanina, N.V.; Vasilyev, P.Y. Optical Coatings Based on CNTs for the Optical Devises and Nanoelectronics. RU Patent 2 355 001 C2, 9 January 2007.
22. Kamanina, N.V.; Vasilyev, P.Y.; Studeonov, V.I. Optical Coating Based on Oriented in the Electric Field CNTs for the Optical Devises, Micro- and Nanoelectronics under the Conditions When the Interface: Solid Substrate-Coating Can Be Eliminated. RU Patent 2 405 177 C2, 23 December 2008.

23. Kamanina, N.V.; Likhomanova, S.V.; Kuzhakov, P.V. Advantages of the surface structuration of KBr materials for spectrometry and sensors. *Sensors* **2018**, *18*, 3013. [CrossRef] [PubMed]
24. Kamanina, N.; Kuzhakov, P.; Kukharchik, A.; Kvashnin, D. A nanostructuring approach for modification of the features of optical materials: Lithium fluoride. In Proceedings of the IOP Conference Series: Materials Science and Engineering, III International Scientific–Practical Conference "Graphene and Related Structures: Synthesis, Production, and Application", Tambov, Russia, 13–15 November 2019; Volume 693, p. 012008. [CrossRef]
25. Plimpton, S. Fast parallel algorithms for short-range molecular dynamics. *J. Comput. Phys.* **1995**, *117*, 1–19. [CrossRef]
26. Daw, M.S.; Baskes, M.I. Semiempirical, quantum mechanical calculation of hydrogen embrittlement in metals. *Phys. Rev. Lett.* **1983**, *50*, 1285–1288. [CrossRef]
27. Tersoff, J. Modeling solid-state chemistry: Interatomic potentials for multicomponent systems. *Phys. Rev. B.* **1989**, *39*, 5566–5568. [CrossRef] [PubMed]
28. Kvashnin, D.G.; Sorokin, P.B. Effect of ultrahigh stiffness of defective graphene from atomistic point of view. *J. Phys. Chem. Lett.* **2015**, *6*, 2384–2387. [CrossRef] [PubMed]

coatings

MDPI

Article

Highly Conductive Mn-Co Spinel Powder Prepared by Cu-Doping Used for Interconnect Protection of SOFC

Zhou Jiang [1,2], Kui Wen [2,*], Chen Song [2], Taikai Liu [2,*], Yong Dong [1], Min Liu [2], Changguang Deng [2], Chunming Deng [2] and Chenghao Yang [3]

1 School of Materials and Energy, Guangdong University of Technology, Guangzhou 510640, China; jiangzhou_2019@163.com (Z.J.); yongdong@gdut.edu.cn (Y.D.)
2 National Engineering Laboratory for Modern Materials Surface Engineering Technology, The Key Lab of Guangdong for Modern Surface Engineering Technology, Guangdong Institute of New Materials, Guangdong Academy of Science, Guangzhou 510651, China; songchen@gdinm.com (C.S.); liumin@gdinm.com (M.L.); dengchangguang@gdinm.com (C.D.); dengchunming@gdinm.com (C.D.)
3 School of Environment and Energy, South China University of Technology, Guangzhou 510006, China; esyangc@scut.edu.cn
* Correspondence: wenkui@gdinm.com (K.W.); liutaikai@gdinm.com (T.L.)

Abstract: Mn-Co Spinel is considered as one of the most promising materials for the interconnect protection of solid oxide fuel cells; however, its conductivity is too low to maintain a high cell performance as compared with cathode materials. Element doping is an effective method to improve the spinel conductivity. In this work, we proposed doping Mn-Co spinel powder with Cu via a solid phase reaction. $Cu_\delta Mn_{1.5-x}Co_{1.5-y}O_4$ with δ = 0.1, 0.2, 0.3, and x + y = δ was obtained. X-ray diffraction (XRD) and thermogravimetry-differential scanning calorimetry (TG-DSC) were used to evaluate the Cu-doping effect. After sintering at 1000 °C for 12 h, the yield exhibited the best crystallinity, density, and element distribution, with a phase composition of $MnCo_2O_4/Cu_xMn_{3-x}O_4$ (x = 1, 1.2, 1.4 or 1.5). X-ray photoelectron spectroscopy (XPS) was used to semi-quantitatively characterize the content changes in element valence states. The areal fraction of Mn^{2+} and Co^{3+} was found to decrease when the sintering duration increased, which was attributed to the decomposition of the $MnCo_2O_4$ phase. Finally, coatings were prepared by atmospheric plasma spraying with doped spinel powders and the raw powder $Mn_{1.5}Co_{1.5}O_4$. It was found that Cu doping can effectively increase the conductivity of Mn-Co spinel coatings from 23 S/cm to 51 S/cm. Although the dopant Cu was found to be enriched on the surface of the coatings after the conductivity measurement, which restrained the doping effect, Cu doping remains a convenient method to significantly promote the conductivity of spinel coatings for SOFC applications.

Keywords: Mn-Co spinel; Cu doping; conductivity; SOFC; interconnect

1. Introduction

Solid oxide fuel cells (SOFC) are high-temperature energy conversion devices that can directly convert the chemical energy of fuel into electrical energy. They mainly consist of a porous anode/cathode, a dense electrolyte, and an interconnect. As an important part, the interconnect provides electrical contact between cells while isolating fuel gas from air. Ferritic stainless steel (FSS) has been widely used as interconnect material due to its outstanding oxidation resistance and suitable thermal expansion coefficient [1–3]. However, at elevated temperature, the Cr of FSS is oxidized to form Cr_2O_3, CrO_3, or even $CrO_2(OH)_2$. These substances can accumulate at the cathode and react with cathode materials, forming a low active phase and deteriorating cell performance. At present, the above problems are mainly solved by introducing protective coatings [4,5].

$(Mn,Co)_3O_4$ spinel coating is often used to diminish the evaporation of Cr and thus reduce the performance degradation and prolong the service life of the stack. However, the theoretical conductivity of $(Mn,Co)_3O_4$ spinel at elevated temperature is only 60 S/cm, which is much lower than that of the cathode material, such as LSCF > 200 S/cm at 600–800 °C [6–9]. Many efforts have been made to improve the conductivity of Mn-Co spinel coating. Fe doping can reduce the growth of Cr_2O_3 film and improve the oxidation resistance, but this reduces the conductivity of spinel coatings [10–13]. Ce doping can enhance the adhesion and improve the interfacial stability of oxide film, but the effects depend on the substrate materials [14–16]. Cu doping can improve the conductivity and adjust the thermal expansion coefficient of spinel coatings [17–19].

For promoting the conductivity of Mn-Co spinel coatings, in this work, we prepared Cu-doped Mn-Co spinel powder via the solid-phase reaction method. The doped powder was denoted as $Cu_\delta Mn_{1.5-x}Co_{1.5-y}O_4$ (δ = 0.1, 0.2, 0.3, x + y = δ). The first part of this work was to study the optimal doping amount of Cu, and then the doping conditions were studied to improve the doping effect. Finally, the doped powder was used to prepare coatings on FSS substrate to evaluate the doping effect using a plasma spray.

2. Experiment

2.1. Powder Preparation and Characterization

The raw materials, as shown in Figure 1, were Cu powders (Tijo, Changsha, China) with D_{50} = 21.58 μm, and $Mn_{1.5}Co_{1.5}O_4$ powder with D_{50} = 19.2 μm (Terio, Qingdao, China). Three groups of doping powder were prepared with controlled Cu-doping as shown in Table 1. The raw powders were mechanically mixed in a roller mixer at a speed of 600 r/min for 20 h. After mixing, the powders were transferred into a muff furnace for solid-phase reaction sintering, and then the sintered powders were sieved to maintain the particle size below 50 μm. The phase structure of the sieved powders was checked by X-ray diffraction (XRD, Bruker D8 Advance, Bruker Cop., Karlsruhe, Germany) in the range of 10–90° (0.01°/step, 0.167 s/step). The micromorphology and element distribution of the sieved powders were characterized by a scanning electron microscope (SEM, Nova-Nano-450, FEI, Hillsboro, OR, USA) and energy dispersive spectrometer (EDS, XLT TEM-SDD, AMETEK EDAX, Philadelphia, PA, USA). The solid-phase sintering process was investigated using a thermogravimetry-differential scanning calorimetry (TG-DSC, STA-449-F5, NETZSCH, Selb, Germany) test with a heating rate of 5 °C/min in air. The valence states of the sintered powders were analyzed by X-ray photoelectron spectroscopy (XPS, Thermo Fisher Scientific Nexsa, Thermo Fisher Scientific Inc., Waltham, MA, USA), and then the content changes in the element valence states were characterized semi-quantitatively. The XPS spectra were obtained with monochromatized Al Kα radiation. All spectra were recorded at a take-off angle of 90°. The carbon C 1s peak with a binding energy of 284.5 eV was used for calibration.

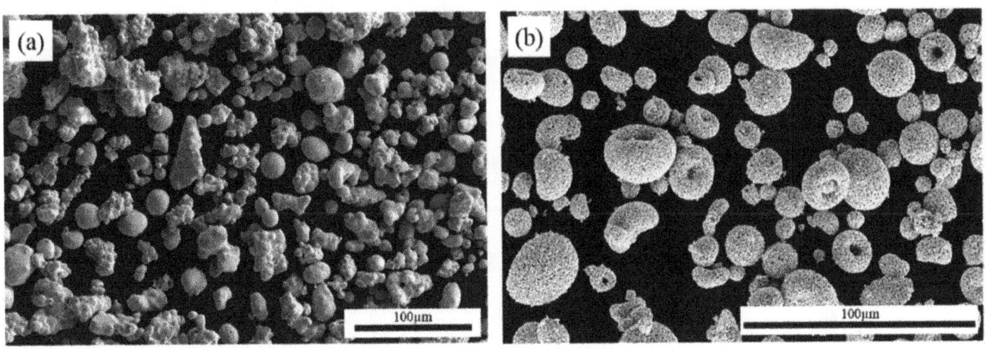

Figure 1. The morphology of Cu powder (**a**) and $Mn_{1.5}Co_{1.5}O_4$ powder (**b**).

Table 1. The raw material ratio to prepare $Cu_{\delta}Mn_{1.5-x}Co_{1.5-y}O_4$ powders.

Mass Percentage (wt%)	δ = 0.1	δ = 0.2	δ = 0.3
Cu Powder	2.7	5.4	8.0
$Mn_{1.5}Co_{1.5}O_4$ Power	97.3	94.6	92.0

2.2. Coating Preparation

Al_2O_3 sheets (φ 18.55 mm × 1.65 mm) were used as substrates. $Mn_{1.5}Co_{1.5}O_4$ and doped $Cu_{0.3}Mn_{1.5-x}Co_{1.5-y}O_4$ powders were used as spraying feedstocks. Atmospheric plasma spraying equipment (model CLWI-091, GTV, Luckenbach, Germany) was used to prepare the spinel coating. The spraying process parameters are shown in Table 2. Before spraying, substrates were ultrasonically cleaned with alcohol to remove any oil and impurities on the surface.

Table 2. Spray parameters to prepare Cu-doped Mn-Co spinel coatings.

Gas Flow (NL/min)		Current (A)	Gun Speed (mm/s)	Powder Feeding Rate (g/min)	Spray Distance (mm)
Ar/72	H_2/8	620	800	15	110

2.3. Coating Characterization

The sectional morphology, density, and element distribution of the spinel coating were analyzed by SEM and EDS. The phase composition of the coating was characterized by XRD. The conductivity of coatings was determined by the four-probe method. The conductivity test setup is shown in Figure 2. Testing samples were placed in a tubular furnace, which was heated up to 800 °C at a rate of 5 °C/min. A constant current of 0.2 A was provided between I^+ and I^- through silver lines, and the corresponding voltage between U^+ and U^- was recorded. The coating conductivity (ε, S/cm) was obtained as $\varepsilon = 1 \div (C \times R \times W)$, where C denotes the wiring coefficient, R denotes the calculated resistance (Ω), and W denotes the thickness of the coating (cm).

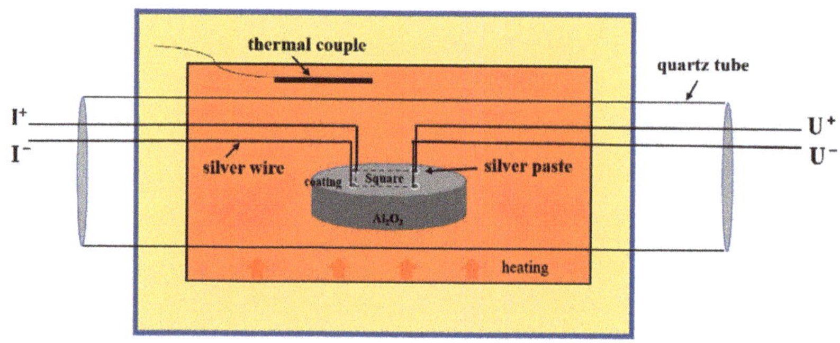

Figure 2. Illustration of the conductivity measurements setup.

3. Results and Discussion

3.1. Phases and Structures

Figure 3 shows the XRD patterns of $Cu_{\delta}Mn_{1.5-x}Co_{1.5-y}O_4$ (δ = 0, 0.1, 0.2, 0.3, x + y = δ) powders sintered at 900 °C for 12 h. $Cu_{\delta}Mn_{1.5-x}Co_{1.5-y}O_4$ (δ = 0.1, 0.2, 0.3) powders were composed of CuO, $MnCo_2O_4$, $Cu_xMn_{3-x}O_4$ (x = 1, 1.2, 1.4, or 1.5), and $(Co,Mn)(Co,Mn)_2O_4$. $(Co,Mn)(Co,Mn)_2O_4$ is a mixture of $MnCo_2O_4$ and $CoMn_2O_4$. Cu-doping can promote the formation of the cubic spinel phase. After increasing the amount of Cu-doping, the diffraction peaks were found to shift slightly toward the right because of the replacement of Co^{2+} (0.074 nm) with Cu^{2+} (0.073 nm), and the diffraction intensity of $(Co,Mn)(Co,Mn)_2O_4$ and CuO was found to decrease, which indicates the doping reactions had taken place. When both CuO and $CoMn_2O_4$ were present in the coating, the uniformity and conductivity of the powder were reduced [18]. It was found that the $(Co,Mn)(Co,Mn)_2O_4$ and CuO of $Cu_{0.3}Mn_{1.5-x}Co_{1.5-y}O_4$ powder exhibited the lowest diffraction intensity and the best crystallinity of the spinel phase.

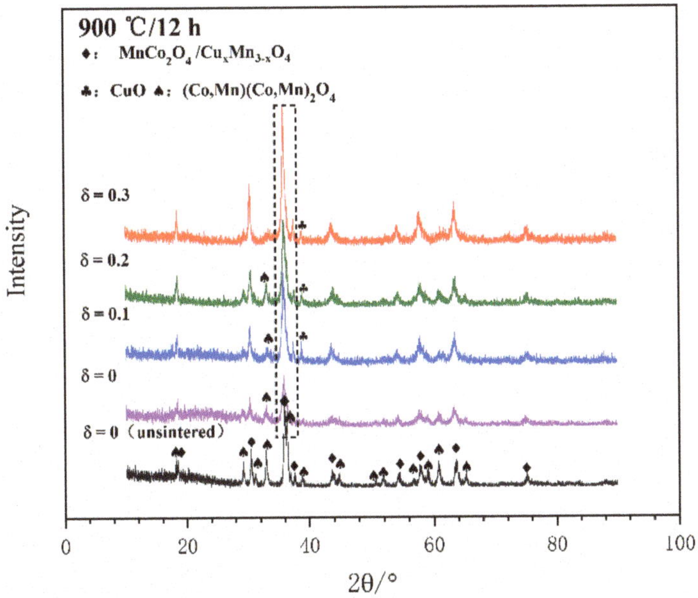

Figure 3. XRD patterns of $Mn_{1.5}Co_{1.5}O_4$ powder and $Cu_\delta Mn_{1.5-0.5x}Co_{1.5-0.5x}O_4$ powder.

3.2. Solid Phase Reactions

$Cu_\delta Mn_{1.5-x}Co_{1.5-y}O_4$ (δ = 0, 0.1, 0.2, 0.3) powders were tested by TG-DSC to evaluate the physical and chemical changes during sintering. According to the TG curve in Figure 4a and the subsequent XRD analysis (Figure 5), it was found that the powders gained some weight at 500–800 °C, which was mainly attributed to the oxidization of Cu to CuO. However, at 800–1000 °C, the powder weights were found to decrease slightly and the CuO had disappeared. At 1034 °C, the powder weight was slightly decreased again, which might be related to the decomposition of the spinel phase. Then, the powder was found to lose weight violently between 1200–1300 °C, which indicates that the spinel phases were decomposed. As shown in Figure 4b, the commercial $Mn_{1.5}Co_{1.5}O_4$ powder had no endothermic peak at 1034 °C, while $Cu_\delta Mn_{1.5-x}Co_{1.5-y}O_4$ (δ = 0.1, 0.2, 0.3) powder had an endothermic peak. According to public reports [9,13,20,21], the formation of $Cu_xMn_{3-x}O_4$ and $MnCo_2O_4$ is an exothermic process. Since the $MnCo_2O_4$ phase of the commercial $Mn_{1.5}Co_{1.5}O_4$ powder showed no significant decomposition at 1034 °C, it was inferred that the endothermic peak at 1034 °C was the result of the decomposition of $Cu_xMn_{3-x}O_4$. Furthermore, in the range 1100–1300 °C, after increasing the amount of Cu, the TG curve shifted toward the left, and the stable temperature of the spinel phases decreased. The DSC curves were found to gradually flatten and continuously exhausted heat. As shown in Figure 4c,d, the $Cu_{0.3}Mn_{1.5-x}Co_{1.5-y}O_4$ powder exhibited the best crystallization. In conclusion, $Cu_xMn_{3-x}O_4$ spinel was successfully synthesized between 900 °C and 1034 °C. The powders with δ = 0.3 had the best crystallinity of spinel phases and consisted of more conductive spinel phase.

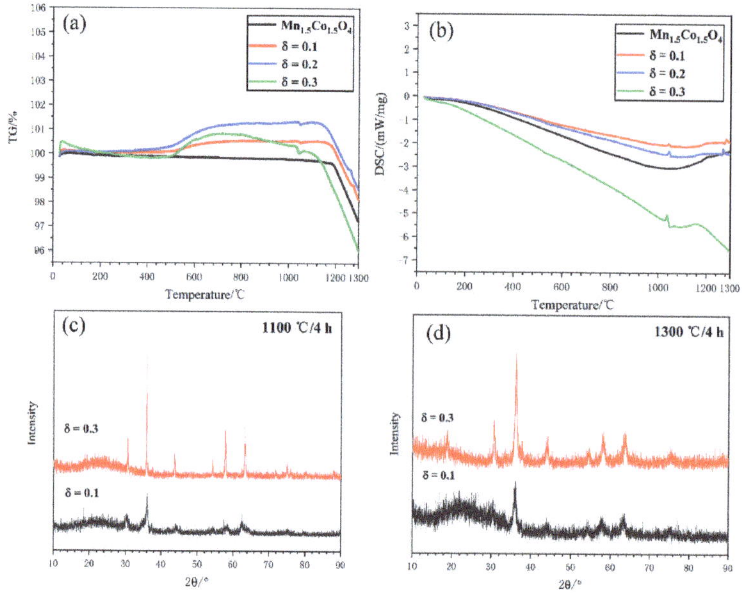

Figure 4. (**a**) TG curves of $Cu_{\delta}Mn_{1.5-x}Co_{1.5-y}O_4$ powders; (**b**) DSC curve; (**c**) XRD patterns of powder sintered at 1100 °C for 4 h with δ = 0.1 and 0.3; (**d**) XRD patterns of powder sintered at 1300 °C for 4 h with δ = 0.1 and 0.3.

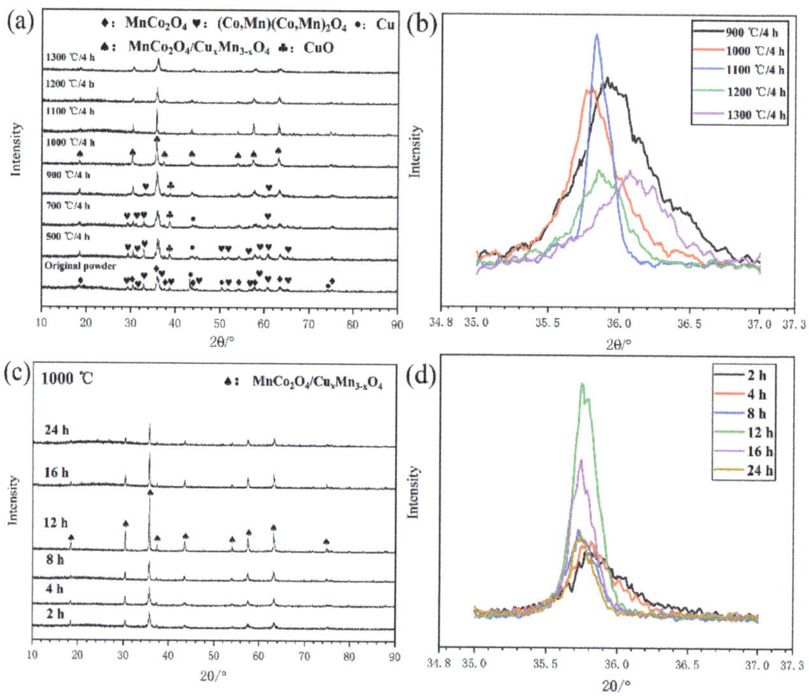

Figure 5. XRD patterns of doped powders sintered under different conditions: (**a**,**b**) 500–1300 °C for 4 h; (**c**,**d**) 1000 °C for 2–24 h.

3.3. Sintering Conditions

$Cu_{0.3}Mn_{1.5-x}Co_{1.5-y}O_4$ powders were prepared at controlled sintering temperatures and durations. Cu was introduced into the spinel lattice by sintering. The more Cu that entered the spinel lattice, the better the conductivity of the coating. The phase composition, crystallization, and surface morphology of the prepared $Cu_{0.3}Mn_{1.5-x}Co_{1.5-y}O_4$ powders were specifically investigated. According to the XRD patterns of the powders sintered at 500–1300 °C for 4 h, it was found that, with the increase in sintering temperature, the abundance of CuO and $(Co,Mn)(Co,Mn)_2O_4$ phases reduced, while the abundance of $MnCo_2O_4/Cu_xMn_{3-x}O_4$ spinel phases increased. The crystallinity of the spinel phases increased firstly and then decreased as the temperature increased; thus, the best crystallization temperature was between 1000–1100 °C, as shown in Figure 5. According to the morphology of the sintered powders, the powder density was found to increase when the temperature increased from 900 °C to 1300 °C. After sintering between 1100–1300 °C, the powder was found to be severely agglomerated, as shown in Figure 6, while at 1000 °C, the powder porosity was reduced.

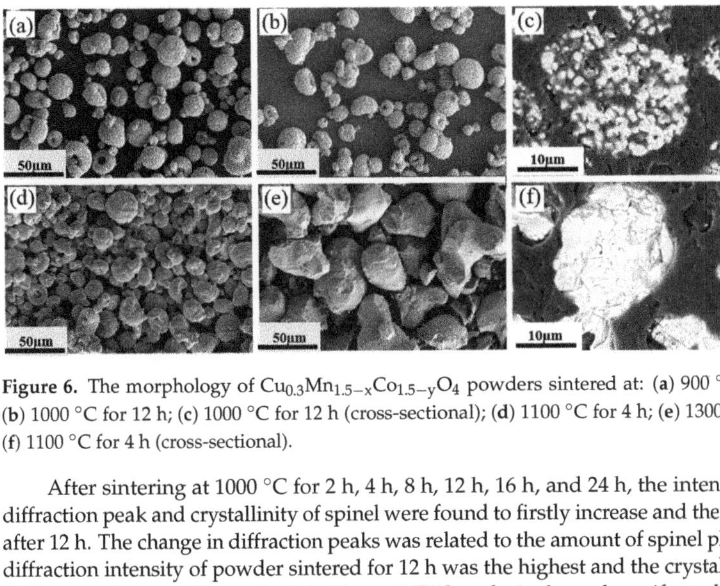

Figure 6. The morphology of $Cu_{0.3}Mn_{1.5-x}Co_{1.5-y}O_4$ powders sintered at: (**a**) 900 °C for 12 h; (**b**) 1000 °C for 12 h; (**c**) 1000 °C for 12 h (cross-sectional); (**d**) 1100 °C for 4 h; (**e**) 1300 °C for 4 h; (**f**) 1100 °C for 4 h (cross-sectional).

After sintering at 1000 °C for 2 h, 4 h, 8 h, 12 h, 16 h, and 24 h, the intensity of the diffraction peak and crystallinity of spinel were found to firstly increase and then decrease after 12 h. The change in diffraction peaks was related to the amount of spinel phases. The diffraction intensity of powder sintered for 12 h was the highest and the crystallinity was the best, as shown in Figure 5c,d. In Figure 7, EDS analysis showed a uniform distribution of Cu, which indicates that the Cu has been sucessfully doped into $Mn_{1.5}Co_{1.5}O_4$ spinel powders. In conclusion, 1000 °C/12 h was the best condition for Cu doping.

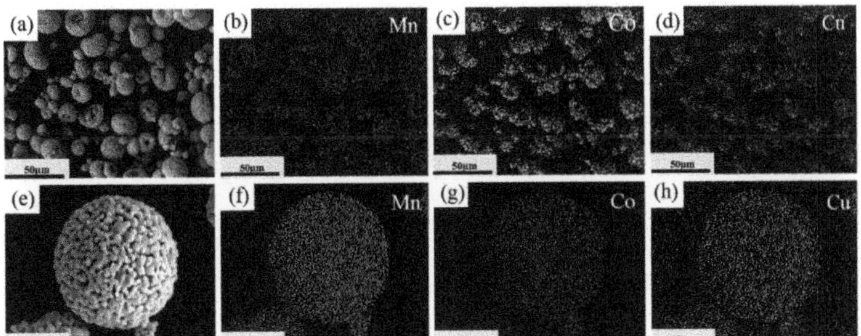

Figure 7. The morphology and elements distribution of $Cu_{0.3}Mn_{1.5-x}Co_{1.5-y}O_4$ powders sintered at 1000 °C for 12 h (**a–d**) low magnification; (**e–h**) high magnification.

3.4. Stability of Doped Powders

As Cu-Mn spinel is less stable than Mn-Co spinel at elevated temperature, when doping Mn-Co spinel with Cu to improve the conductivity, the stability of the obtained spinel is reduced. After being sintered at 1000 °C for 4 h, 12 h, and 24 h to allow the doping, the $Cu_{0.3}Mn_{1.5-x}Co_{1.5-y}O_4$ powder was heated to 800 °C and maintained in that state for 12 h to evaluate the stability. In Figure 8a, the XRD patterns before and after the heat treatment are presented. The $MnCo_2O_4/Cu_xMn_{3-x}O_4$ phase was detected after the heat treatment, but the diffraction intensity was significantly diminished. TG and DSC tests were performed from room temperature to 800 °C with a ramping rate of 5 °C/min, and a preservation stage of 12 h, in order to study the stability of the doped powders. It was found that during ramping, the DSC plots showed evident exothermic behavior. However, during the following preservation stage, both the TG and DSC curves exhibited a linear behavior in which no endothermic and exothermic peak was visible, as shown in Figure 8b–d. This indicates that the doped powder had a good stability during the preservation process. For the $Cu_xMn_{3-x}O_4$ powder, when x = 1.2, 1.4, 1.5, the CuO and spinel phase coexisted at 800 °C. However, according to the XRD and DSC patterns, it can be seen that the decomposition of $Cu_xMn_{3-x}O_4$ spinel did not occur and CuO was not formed. Therefore, the $Cu_xMn_{3-x}O_4$ phase was mainly composed of the $CuMn_2O_4$ phase. In conclusion, the $MnCo_2O_4/CuMn_2O_4$ phase in the powder can exist stably at 800 °C.

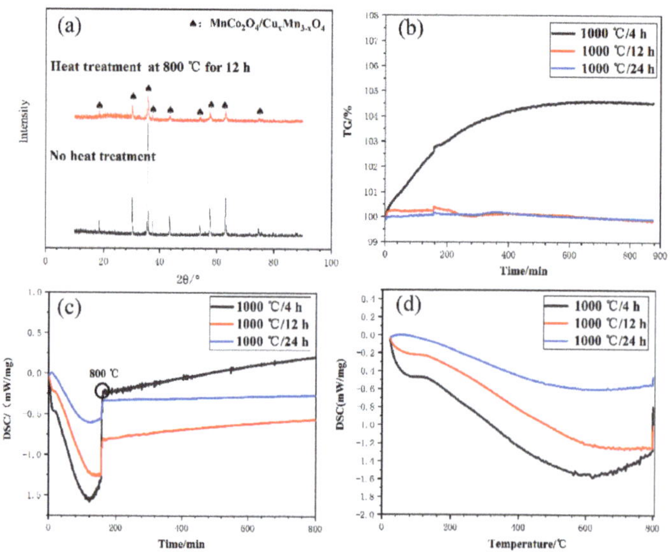

Figure 8. (**a**) XRD patterns before and after heating at 800 °C of powder sintered at 1000 °C for 12 h; (**b**,**c**) TG-DSC curves of powders sintered at 1000 °C for 4 h, 12 h, and 24 h; (**d**) DSC curves during the heating process.

3.5. Elemental Composition

To determine the elemental composition and valence states of $Cu_{0.3}Mn_{1.5-x}Co_{1.5-y}O_4$ powders, XPS tests were carried out. As shown in Figure 9a, Co, Mn, Cu, O, and C were identified in the coating. According to the Co 2p spectrum in Figure 9b, the low-energy peaks (780.6 and 796.1 eV) and high-energy peaks (781.6 and 797.4 eV) in the Co $2P_{3/2}$ and $2P_{1/2}$ regions were ascribed to Co^{3+} and Co^{2+}, respectively. In addition, the peaks with binding energies of 786.6 eV and 803.6 eV were identified as the satellite peaks of Co^{2+} and Co^{3+} [22–24]. As illustrated in Figure 9c, the peaks at 641.5, 642.3, and 642.8 eV corresponded to Mn^{2+}, Mn^{3+}, and Mn^{4+}, respectively [25–27]. In Figure 9d, the peaks at 945 eV were attributed to Cu^{2+} satellites, while those at 930.8 and 933.4 eV were evidence of the presence of Cu^+ and Cu^{2+} in the Cu $2P_{3/2}$ orbital peaks. Peaks at 953.5 and 950.6 eV can be assigned to the existence of Cu^{2+} and Cu^+, which were observed in the Cu $2P_{1/2}$ orbital [28–30]. The peaks at 530.4 eV and 531.9 eV, as shown in Figure 9e, corresponded to the oxygen-metal bond (metal = Mn, Co) and O-H bond in the spinel phase, respectively [3].

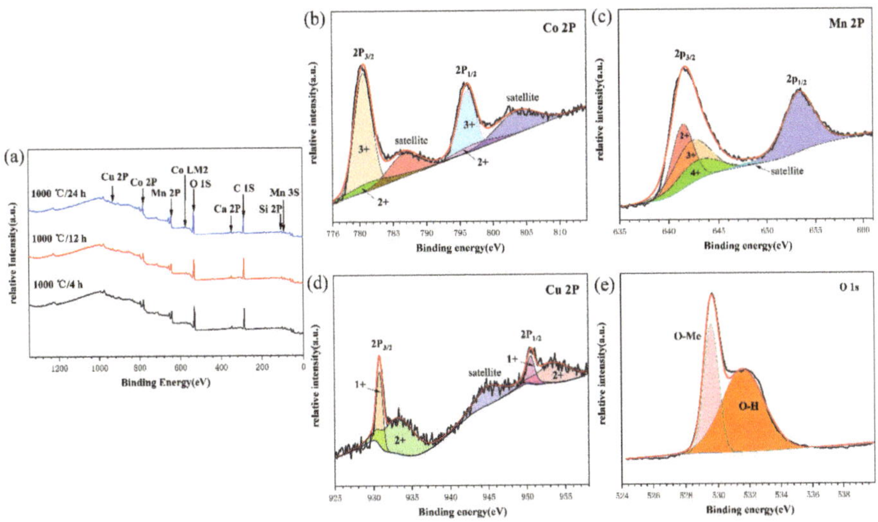

Figure 9. XPS spectra of $Cu_{0.3}Mn_{1.5-x}Co_{1.5-y}O_4$ powders sintered at 1000 °C: (**a**) survey; (**b**) Co 2P of 1000 °C/12 h sample; (**c**) Mn 2P of 1000 °C/12 h sample; (**d**) Cu 2P of 1000 °C/12 h sample; (**e**) O 1s of 1000 °C/12 h sample.

According to public reports [10,31,32], various valences exist in Mn-Co-Cu spinel. By performing a deconvolution of the XPS spectra, the relative proportion of different valence states of elements can be obtained. The relative proportion is listed in Table 3. It was found that the relative proportion of Mn^{3+} increased to 41.3% when sintered for 12 h and then decreased 33.6% after 24 h of preservation. The areal fraction of Mn^{2+} and Co^{3+} increased first and then decreased when the sintering duration increased. The increase in Mn^{3+} indicated an increase in the $CuMn_2O_4$ phase, and the decrease in Mn^{2+} and Co^{3+} was related to the decomposition of the $MnCo_2O_4$ phase. Moreover, the increase in Mn^{4+} indicated an increase in MnO_2, which diminished the conductivity of the prepared powders or coatings. According to Table 3 the powder prepared at 1000 °C and sintered for 12 h gave the most $MnCo_2O_4$ and $CuMn_2O_4$.

Table 3. Areal fractions of Mn and Co in the powders prepared under different conditions.

Element State	Areal Fraction (%)			
	1000 °C/4 h	1000 °C/8 h	1000 °C/12 h	1000 °C/24 h
Mn^{2+}	32.3	46.2	36.7	34.2
Mn^{3+}	31.7	33.2	41.3	33.6
Mn^{4+}	36.0	20.6	22.0	32.2
Co^{2+}	23.6	20.9	17.7	26.3
Co^{3+}	76.4	79.1	82.3	73.7

3.6. Coatings Conductivity

To evaluate the doping effect, commercial $Mn_{1.5}Co_{1.5}O_4$ and $Cu_{0.3}Mn_{1.5-x}Co_{1.5-y}O_4$ powder (sintered at 1000 °C and maintained for 8 h, 12 h, and 24 h, respectively) were sprayed on Al_2O_3 sheets with an atmospheric plasma spray; the parameters are provided in Table 1. During spraying, the spinel powder was injected into the plasma plume, the temperature of which was as high as 15,000 °C, then the injected powder was heated, melted, accelerated, and impacted on the substrate to form coatings. Deoxidization and decomposition of the spinel powder occurred during spraying; therefore, low active phases such as MnO_2 and CoO were generated and left in the coating. As a result of the very short spraying time, only a small part of the powder was decomposed, and most of the powder retained the spinel structure. Before the conductivity measurement, the sample was heated to 800 °C and maintained for 12 h to allow the densification of the coating and cure the decomposed phases. As shown in Figure 10, after the heat treatment, significant signals of $MnCo_2O_4/CuMn_2O_4$ were detected, which indicates a transference of low conductive phases to the spinel phase. For this reason, the measured coating conductivity increased with the temperature during preservation with a duration up to 50 h, as shown in Figure 11. The conductivity of the coating prepared with powder sintered for 12 h was about 51 S/cm after 47 h of preservation. For the coating prepared with powder sintered at 1000 °C for 8 h and 24 h, the conductivity was 45 S/cm and 49.2 S/cm, respectively, which are more than double that of the $Mn_{1.5}Co_{1.5}O_4$ coating, 23 S/cm after 50 h of preservation. Therefore, by applying Cu doping, the conductivity of Mn-Co spinel coatings can be significantly improved. Moreover, according to Figures 10 and 11, the coating prepared with powder sintered at 1000 °C for 8 h exhibited lower crystallinity and less spinel content, resulting in a coating conductivity lower than that of the 1000 °C/12 h and 24 h samples.

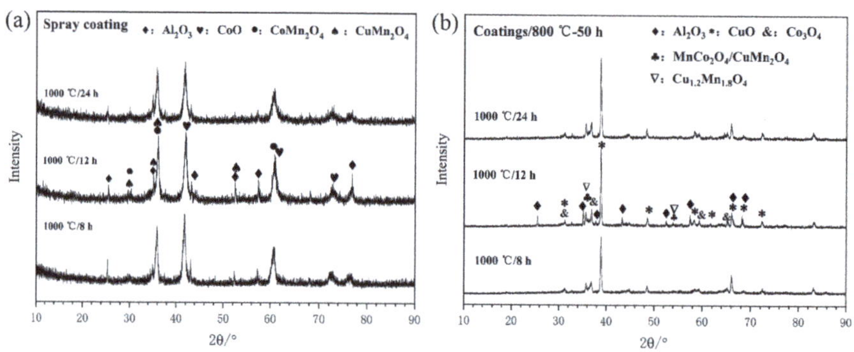

Figure 10. XRD patterns of coatings prepared with sintering powder for different times: (**a**) spray state; (**b**) coatings at 800 °C for 50 h.

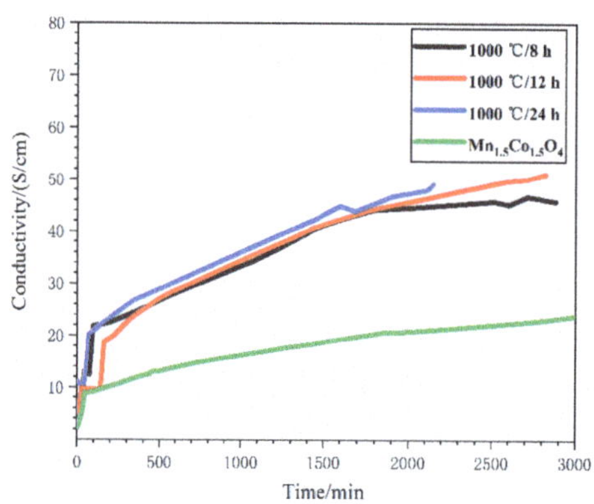

Figure 11. Coatings conductivity prepared from commercial spinel powder and $Cu_{0.3}Mn_{1.5-x}Co_{1.5-y}O_4$.

The sectional morphology and element distribution before and after the conductivity measurement of the best coating (1000 °C for 12 h) are provided in Figure 12. After the conductivity measurement, the coating was significantly densified, and no large pores or cracks were visible on the cross-section. The distribution of Mn and Co exhibited no evident variation, while Cu was found to be enriched on the coating surface. The enriched Cu on the surface was then oxidized to CuO at 800 °C, which limited the conductivity of the coatings. Despite the existence of CuO on the surface restraining the doping effect, Cu doping remains a convenient method to significantly promote the conductivity of spinel coatings for SOFC applications.

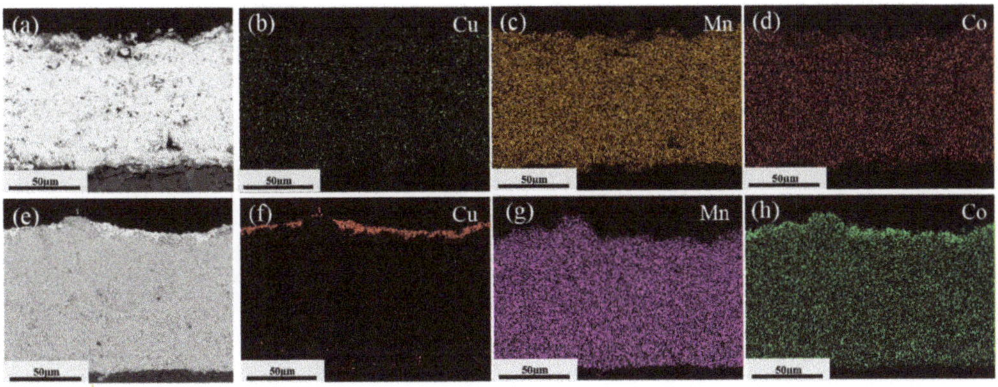

Figure 12. Morphologies and compositions of $Cu_{0.3}Mn_{1.5-x}Co_{1.5-y}O_4$ coating: (**a–d**) before the conductivity measurement; (**e–h**) after the conductivity measurement.

4. Conclusions

The powder $Cu_\delta Mn_{1.5-x}Co_{1.5-y}O_4$ ($x + y = \delta$) with $\delta = 0.3$ showed a better performance than the powder with $\delta = 0.1$ or 0.2. The results of XRD and TG-DSC indicated that increasing the dopant amount resulted in an enhanced doping effect. With $\delta = 0.3$, the powder had a better crystallinity and a more conductive spinel phase.

The best doping condition for $Cu_{0.3}Mn_{1.5-x}Co_{1.5-y}O_4$ powders was 1000 °C for 12 h. Under this condition, the crystallization of the spinel phase ($MnCo_2O_4$ and $Cu_xMn_{3-x}O_4$) was the best and the powder was moderately dense. The dopant Cu was found to be evenly distributed in the powders. The $Cu_xMn_{3-x}O_4$ ($x = 1$, 1.2, 1.4 or 1.5) was found to be mainly composed of $CuMn_2O_4$ and was sufficiently stable at elevated temperatures. According to the XPS results, the powder prepared at 1000 °C for 12 h consisted of the most $MnCo_2O_4$ and $CuMn_2O_4$. When the preservation stage was conducted at 1000 °C with a duration of less than 12 h or more than 12 h, less $MnCo_2O_4$ and $Cu_xMn_{3-x}O_4$ phases were detected.

Coatings were prepared with doped powders using an atmospheric plasma spray. After preservation at 800 °C for a duration of up to 50 h, the measured conductivity of the prepared coatings reached 45 S/cm, 51 S/cm, and 49.2 S/cm, respectively, i.e., two times more than that of the $Mn_{1.5}Co_{1.5}O_4$ coating (23 S/cm). Although Cu was enriched on the coating surface after the conductivity measurement, Cu-doping is still an effective method to improve the conductivity of spinel coatings used for interconnect protection in SOFC.

Author Contributions: Formal analysis, Z.J.; investigation, Z.J. and C.S.; methodology, K.W. and T.L.; project administration, M.L.; resources, C.D. (Changguang Deng); supervision, C.D. (Chunming Deng) and Y.D.; writing—original draft, Z.J.; writing—review and editing, T.L. and C.Y. All authors have read and agreed to the published version of the manuscript.

Funding: This research was funded by the National Key R&D Program of China grant number 2018YFB1502603, Guangdong Academy of Science grant number 2019GDASYL-0102007, the Foreign Cooperation Platform Project of Guangdong Province grant number 2020A050519001, and Guangzhou Major Projects of Industry-University-Research (IUR) Collaborative Innovation "Surface Treatment and Repair for Key Components of Industrial Gas Turbine (IGT)".

Institutional Review Board Statement: Not applicable.

Informed Consent Statement: Not applicable.

Data Availability Statement: The data presented in this study are available on request from the corresponding author.

Conflicts of Interest: The authors declare no conflict of interest. The founding sponsors had no role in the design of the study; in the collection, analyses, or interpretation of data; in the writing of the manuscript, and in the decision to publish the results.

References

1. Shong, W.J.; Liu, C.K.; Chen, C.Y.; Peng, C.C.; Tu, H.J.; Fey, G.T.K.; Lee, R.Y.; Kao, H.M. Effects of lanthanum-based perovskit coatings on the formation of oxide scale for ferritic SOFC interconnect. *Mater. Chem. Phys.* **2011**, *127*, 45–50. [CrossRef]
2. Stanislowski, M.; Froitzheim, J.; Niewolak, L.; Quadakkers, W.J.; Hilpert, K.; Markus, T.; Singheiser, L. Reduction of chromiun vaporization from SOFC interconnectors by highly effective coatings. *J. Power Sources* **2007**, *164*, 578–589. [CrossRef]
3. Zou, J.; Song, C.; Wen, K.; Liu, T.; Deng, C.; Liu, M.; Yang, C. The Microstructure and Conductivity Evolution of Plasma-Spraye $(Mn, Co)_3O_4$ Spinel Coatings during Conductivity Measurements at Elevated Temperature. *Coatings* **2021**, *11*, 533. [CrossRef]
4. Grünwald, N.; Sohn, Y.J.; Yin, X.; Menzler, N.H.; Guillon, O.; Vaßen, R. Microstructure and phase evolution of atmospheri plasma sprayed Mn-Co-Fe oxide protection layers for solid oxide fuel cells. *J. Eur. Ceram. Soc.* **2019**, *39*, 449–460. [CrossRef]
5. Xin, X.; Zhu, Q.; Liu, Y. Conductive protective coating with heat oxygen-resistance for solid oxide fuel cell (SOFC) allo interconnect. *Surf. Technol.* **2019**, *48*, 22–29. [CrossRef]
6. Jiang, S.P. Development of lanthanum strontium cobalt ferrite perovskite electrodes of solid oxide fuel cells-A review. *Int. Hydrogen Energy.* **2019**, *44*, 7448–7493. [CrossRef]
7. Tan, K.H.; Rahman, H.A.; Taib, H. Coating layer and influence of transition metal for ferritic stainless steel interconnector soli oxide fuel cell: A review. *Int. J. Hydrogen Energy* **2019**, *44*, 30591–30605. [CrossRef]
8. Mah, J.C.W.; Muchtar, A.; Somalu, M.R.; Ghazali, M.J. Metallic interconnects for solid oxide fuel cell: A review on protectiv coating and deposition techniques. *Int. J. Hydrogen Energy* **2017**, *42*, 9219–9229. [CrossRef]
9. Waluyo, N.S.; Park, S.S.; Song, R.H.; Lee, S.-B.; Lim, T.-H.; Hong, J.-E.; Ryu, K.H.; Bin Im, W.; Lee, J.-W. Protective coating base on manganese–copper oxide for solid oxide fuel cell interconnects: Plasma spray coating and performance evaluation. *Ceram. In* **2018**, *44*, 11576–11581. [CrossRef]
10. Talic, B.; Hendriksen, P.V.; Wiik, K.; Lein, H.L. Thermal expansion and electrical conductivity of Fe and Cu doped $MnCo_2O$ spinel. *Solid State Ion.* **2018**, *326*, 90–99. [CrossRef]
11. Talic, B.; Hendriksen, P.V.; Wiik, K.; Lein, H.L. Diffusion couple study of the interaction between Cr_2O_3 and $MnCo_2O_4$ dope with Fe and Cu. *Solid State Ion.* **2019**, *332*, 16–24. [CrossRef]
12. Puranen, J.; Pihlatie, M.; Lagerbom, J.; Bolelli, G.; Laakso, J.; Hyvärinen, L.; Kylmälahti, M.; Himanen, O.; Kiviaho, J.; Lusvarghi, L et al. Post-mortem evaluation of oxidized atmospheric plasma sprayed Mn–Co–Fe oxide spinel coatings on SOFC interconnector *Int. J. Hydrogen Energy* **2014**, *39*, 17284–17294. [CrossRef]
13. Liu, Y.; Fergus, J.W.; Wang, K.; Cruz, C.D. Crystal Structure, Chemical Stabilities and Electrical Conductivity of Fe-Dope Manganese Cobalt Spinel Oxides for SOFC Interconnect Coatings. *J. Electrochem. Soc.* **2013**, *160*, F1316–F1321. [CrossRef]
14. Zhu, J.H.; Lewis, M.J.; Du, S.W.; Li, Y. CeO_2-doped $(Co,Mn)_3O_4$ coatings for protecting solid oxide fuel cell interconnect alloy *Thin Solid Films* **2015**, *596*, 179–184. [CrossRef]
15. Tseng, H.P.; Yung, T.Y.; Liu, C.K.; Cheng, Y.-N.; Lee, R.-Y. Oxidation characteristics and electrical properties of La- or Ce-dope $MnCo_2O_4$ as protective layer on SUS441 for metallic interconnects in solid oxide fuel cells. *Int. J. Hydrogen Energy* **2020**, *45* 12555–12564. [CrossRef]
16. Gavrilov, N.V.; Ivanov, V.V.; Kamenetskikh, A.S.; Nikonov, A. Investigations of Mn–Co–O and Mn–Co–Y–O coatings deposited b the magnetron sputtering on ferritic stainless steels. *Surf. Coat. Technol.* **2011**, *206*, 1252–1258. [CrossRef]
17. Brylewski, T.; Kruk, A.; Bobruk, M.; Adamczyk, A.; Partyka, J.; Rutkowski, P. Structure and electrical properties of Cu-dope Mn-Co-O spinel prepared via soft chemistry and its application in intermediate-temperature solid oxide fuel cell interconnects. *J Power Sources* **2016**, *333*, 145–155. [CrossRef]
18. Xu, Y.; Wen, Z.; Wang, S.; Wen, T. Cu doped Mn–Co spinel protective coating on ferritic stainless steels for SOFC interconnec applications. *Solid State Ion.* **2011**, *192*, 561–564. [CrossRef]
19. Sabato, A.G.; Molin, S.; Javed, H.; Zanchi, E.; Boccaccini, A.R.; Smeacetto, F. In-situ Cu-doped MnCo-spinel coatings for solid oxide cell interconnects processed by electrophoretic deposition. *Ceram. Int.* **2019**, *45*, 19148–19157. [CrossRef]

20. Wei, P.; Bieringer, M.; Cranswick, L.M.D.; Petric, A. In situ high-temperature X-ray and neutron diffraction of Cu–Mn oxide phases. *J. Mater. Sci.* **2010**, *45*, 1056–1064. [CrossRef]
21. Masi, A.; Bellusci, M.; McPhail, S.J.; Padella, F.; Reale, P.; Hong, J.-E.; Steinberger-Wilckens, R.; Carlini, M. Cu-Mn-Co oxides as protective materials in SOFC technology: The effect of chemical composition on mechanochemical synthesis, sintering behaviour, thermal expansion and electrical conductivity. *J. Eur. Ceram. Soc.* **2017**, *37*, 661–669. [CrossRef]
22. Waskowska, A.; Gerward, L.; Olsen, J.S.; Steenstrup, S.; Talik, E. $CuMn_2O_4$: Properties and the high-pressure induced Jahn-Teller phase transition. *J. Phys. Condens. Matter* **2001**, *13*, 2549–2562. [CrossRef]
23. Hu, Y.Z.; Su, Y.T.; Li, C.X.; Li, C.-J.; Yang, G.-J. Dense $Mn_{1.5}Co_{1.5}O_4$ coatings with excellent long-term stability and electrical performance under the SOFC cathode environment. *Appl. Surf. Sci.* **2020**, *499*, 143726. [CrossRef]
24. Fan, L.Q.; Huang, J.L.; Wang, Y.L.; Geng, C.-L.; Sun, S.-J.; Huang, Y.-F.; Lin, J.-M.; Wu, J.-H. TiO_2 nanotubes supported ultrafine $MnCo_2O_4$ nanoparticles as a superior-performance anode for lithium-ion capacitors. *Int. J. Hydrogen Energy* **2021**, *46*, 35330–35341. [CrossRef]
25. Li, J.; Xiong, S.; Li, X.; Qian, Y. A facile route to synthesize multiporous $MnCo_2O_4$ and $CoMn_2O_4$ spinel quasi-hollow spheres with improved lithium storage properties. *Nanoscale* **2013**, *5*, 2045–2054. [CrossRef]
26. Umezawa, Y.; Reilley, C.N. Effect of argon ion bombardment on metal complexes and oxides studied by x-ray photoelectron spectroscopy. *Anal. Chem.* **1978**, *50*, 1290–1295. [CrossRef]
27. Vepřek, S.; Cocke, D.L.; Kehl, S.; Oswald, H.R. Mechanism of the deactivation of Hopcalite catalysts studied by XPS, ISS, and other techniques. *J. Catal.* **1986**, *100*, 250–263. [CrossRef]
28. Liu, Y.; Song, Z.; Wang, W.; Wang, Z.; Zhang, Y.; Liu, C.; Wang, Y.; Li, A.; Xu, B.; Qi, F. A $CuMn_2O_4/g\text{-}C_3N_4$ catalytic ozonation membrane reactor used for water purification: Membrane fabrication and performance evaluation. *Sep. Purif. Technol.* **2021**, *265*, 118268. [CrossRef]
29. Fang, R.; Liu, F.; Liu, J.; Li, Y. Experimental and theoretical insights into the reaction mechanism of spinel $CuMn_2O_4$ with CO in chemical-looping combustion. *Appl. Surf. Sci.* **2021**, *561*, 150065. [CrossRef]
30. Thaheem, I.; Joh, D.W.; Noh, T.; Lee, K.T. Highly conductive and stable $Mn_{1.35}Co_{1.35}Cu_{0.2}Y_{0.1}O_4$ spinel protective coating on commercial ferritic stainless steels for intermediate-temperature solid oxide fuel cell interconnect applications. *Int. J. Hydrogen Energy* **2019**, *44*, 4293–4303. [CrossRef]
31. Bobruk, M.; Durczak, K.; Dąbek, J.; Dąbek, J.; Brylewski, T. Structure and Electrical Properties of Mn-Cu-O Spinels. *J. Mater. Eng. Perform.* **2017**, *26*, 1598–1604. [CrossRef]
32. Joshi, S.; Petric, A. Nickel substituted $CuMn_2O_4$ spinel coatings for solid oxide fuel cell interconnects. *Int. J. Hydrogen Energy* **2017**, *42*, 5584–5589. [CrossRef]

coatings

MDPI

Article

Validation of Antibacterial Systems for Sustainable Ceramic Tiles

Valeria La Torre [1], Elisa Rambaldi [1], Giulia Masi [2], Silvia Nici [3], Daniele Ghezzi [4], Martina Cappelletti [3] and Maria Chiara Bignozzi [2,*]

[1] Centro Ceramico, Via Martelli 26, 40138 Bologna, Italy; vale.mati91@gmail.com (V.L.T.); rambaldi@centroceramico.it (E.R.)
[2] Department of Civil, Chemical, Environmental and Materials Engineering, University of Bologna, Via Terracini 28, 40131 Bologna, Italy; giulia.masi5@unibo.it
[3] Department of Pharmacy and Biotechnology, University of Bologna, Via Irnerio 42, 40126 Bologna, Italy; silvia.nici@studio.unibo.it (S.N.); martina.cappelletti2@unibo.it (M.C.)
[4] Laboratory of NanoBiotechnology, IRCCS Istituto Ortopedico Rizzoli, Via di Barbiano 1/10, 40136 Bologna, Italy; daniele.ghezzi@ior.it
* Correspondence: maria.bignozzi@unibo.it; Tel.: +39-(05)-12090342

Abstract: Ceramic tiles are bacteriostatic materials; however, the COVID-19 emergency has pushed tile producers to improve surfaces' antibacterial properties. The aim of this work was to validate a silver-based antibacterial treatment applied to porcelain stoneware tiles based on natural and waste materials, thus correlating surface functionalization to tile composition and relevant physical, microstructural, and textural parameters. The treatment was applied before firing, with and without a polymeric primer. Antibacterial activity tests, stain resistance tests, and contact angle measurements were carried out on fired tiles. Further investigations were made by SEM and optical profilometry in order to study the morphological–structural profile of tile surfaces. Results showed strong antibacterial activities for all the functionalized tiles, which were mainly correlated to the morphological and textural parameters of ceramic surfaces, as well as the presence of the polymeric primer.

Keywords: ceramic tile; antibacterial; recycling; porcelain stoneware; silver compound; functionalization; surface treatment

1. Introduction

The current health emergency, caused by the spread of COVID-19, has drawn the attention of the scientific community to the importance of environmental health in relation to the risk of contamination of covering surfaces.

Ceramic tiles are bacteriostatic materials due to their inorganic nature and vitrified surface, which prevent the proliferation of microorganisms and pathogens, and as a result, they are easy to be hygienically cleaned by means of traditional procedures [1,2]. Since ceramic surfaces are susceptible to bacterial contamination in any destination of use, the constant removal of microbial cells from the surface is essential for the sanitization of the environment. In this perspective, the ceramic tile industry aims to process ceramic tiles to imbue the surfaces of tiles with antibacterial properties, already offering a rather wide range of treatments [3–7].

Currently, there are two main types of technologies able to actively control and remove the microbial load affecting surfaces: photocatalytic and non-photocatalytic technologies. They consist of nanostructured coatings based on titanium dioxide (TiO_2) or silver ions (Ag^+) as active agents, respectively. Both of these semiconductor materials are well known to exert a cytotoxic effect toward different species of microorganisms, causing the inhibition of pathogen proliferation [8–10].

In particular, the bactericidal effect seems to be associated with the induction of intracellular biomolecule (DNA, proteins, and lipids) damage, reactive oxygen species

(ROS) generation, and cell wall disruption [10]. Several studies demonstrated that, among the physical states under which metal oxides can interact with microbial cells, metal nanoparticles exhibit high level of antibacterial activity, due to their small size and high surface–volume ratio [11–14].

In general, TiO_2-based photocatalytic technology is limited by lighting conditions (requiring UV illumination); however, in the specific case of ceramic tiles, particular attention should be addressed to the production process. Of all the TiO_2 polymorphs, anatase is known to exert the best photocatalytic performance, secondarily followed by rutile [15,16]. The sintering thermal cycle of the ceramic body includes a temperature range (600–1100 °C) which, critically, involves the transformation of anatase into rutile, with the risk of reducing the photocatalytic efficacy. For this reason, the industrial exploitation of this technology in ceramic tiles is limited to the implementation of TiO_2-based surface treatments [16–19].

Among the non-photocatalytic technologies, silver-based soluble compounds can be applied, thus exploiting the advantageous environmental flexibility of silver, which does not require specific activation conditions. The antibacterial activity is triggered by direct physical contact with pathogens and carried out through the release of bioactive Ag^+ ions. This prerogative ensures the adaptability of silver-based technology to different application techniques, such as the production of ceramic tiles, coatings, surface treatments, or compounds added into the ceramic-body-glaze system [20–22].

Silver-based-soluble-compound antibacterial technology is already known to be effective if applied on materials other than tiles [23,24]. In this regard, several studies show that antibacterial efficacy can be achieved by designing hybrid technologies that utilize both silver and titania [25–28]. In this case, the photocatalytic aspect is based on a surface coating of ceramic materials with photoactive compounds, which, under light irradiation, produce highly ROS. In turn, pollutants or pathogenic species on the surface undergo continuous degradation processes via ROS. Thanks to the presence of Ag and TiO_2, the surface also exhibits antibacterial properties in the dark through the combined action of the photocatalytic coating and Ag.

In this study, a silver-based antibacterial formulation was applied to two types of unglazed porcelain stoneware tiles, prepared with different mixtures.

Porcelain tiles, representing almost 88% of the Italian ceramic tiles product portfolio [29], are generally characterized by a very compact structure, with extremely low porosity, thus ensuring high resistance to abrasion (surface hardness) and mechanical stress (bending and impact), as well as low frost susceptibility. Such features qualify porcelain tiles as suitable for any type of constructive solution [30,31].

Two porcelain stoneware mixtures were intentionally designed: a mix made entirely of natural raw materials (30–40 wt% ball clays, 0–10 wt% kaolin, 40–55 wt% feldspar and 0–15 wt% quartz sand) [32], and a mix based on a high content of recycled material, derived almost exclusively from the waste of the ceramic process itself. A 20 wt% of recycled ceramic material from the total recovery of unfired and fired scraps, including grinding and lapping fine powders, was used [33–35]. The mixture based on recycled raw materials was specifically developed in order to avoid modifying the settings of a real industrial ceramic production cycle.

The novelty of this work is the evaluation of the effectiveness of the investigated silver-based antibacterial treatment—applied before sintering—on the surfaces of traditional and sustainable porcelain stoneware tiles. The silver-based functionalizing treatment was applied using a screen-printing technique, with and without the addition of a polymeric primer to promote the adhesion between the ceramic substrate and the silver coating.

The effect of regular cleaning procedures (cleanability) on the silver-based treatment was assessed through a comprehensive surface analysis investigation (compositional, textural, and/or microstructural characteristics).

2. Materials and Methods

Industrial spry-dried powders of natural and recycled raw materials were used to prepare traditional and high-recycled-content mixtures, named G and R, respectively. Their chemical composition, determined by inductively-coupled plasma emission spectroscopy (ICP-OES Perkin Elmer Optima 3200XL, Waltham, MA, USA), is reported in Tables 1 and 2.

Table 1. Chemical composition of ceramic mixtures G and R (oxides wt%).

Components	G (wt%)	R (wt%)
LOI	2.95	4.45
SiO_2	79.0	75.0
Al_2O_3	13.0	15.0
TiO_2	0.23	0.55
Fe_2O_3	0.16	0.76
CaO	1.27	0.47
MgO	0.02	0.12
K_2O	0.99	2.03
Na_2O	1.77	0.79
SO_3	0.10	0.20
ZrO_2	0.40	0.07

Table 2. Heavy metals in chemical composition of ceramic mixtures G and R (oxides mg/kg).

Components	G (mg/kg)	R (mg/kg)
CdO	12	13
Cr_2O_3	27	67
CuO	8	6
NiO	<5	<5
MnO_2	20	121
ZnO	<5	111
V_2O_5	39	74
CoO	<5	<5
BaO	103	282
PbO	<20	<20

An aqueous solution enriched with metallic silver (2.9–4.9 wt%), characterized by an alkaline pH (8.3–9.3) and low density (1.05–1.10 g/cm^3), was used as the antibacterial treatment. The solution was industrially applied on 25×25 cm^2 ceramic supports before firing, at a concentration of 90–91 g/m^2. Firing was performed at 1250 °C for 39 min total duration (cold to cold) for all tiles. After firing, specimens of 5×5 cm^2 were cut to be tested.

The tested samples were identified using acronyms referring to the fired untreated ceramic tiles (G and R) and the silver-based antibacterial treatment (A). A further code denotes the application of the treatment, which was directly deposited by serigraphic technique onto the ceramic support, as without (A1 samples) or with (A2 samples) a polymeric primer. The latter was an organic suspension with pH in the range of 6.2–7.2 and a density of 1.00–1.05 g/cm^3. The primer was sprayed onto the ceramic surface (at a concentration of about 50 g/m^2) just before the application of the antibacterial treatment. The schema of the two antibacterial systems is shown in Figure 1.

The antibacterial activity of the tiles was assessed against *Escherichia* (*E.*) *coli* strain ATCC8739, mostly following the procedure described in the standard ISO 21196:2011 [36]. Briefly, after sterilization in 70% ethanol, the tiles (with and without silver, i.e., the control sample) were inoculated with 400 μL of bacterial culture, grown up to a final concentration of ~1×10^6 CFU/mL, in nutrient broth (NB), under shaking (150 rpm), at 37 °C. A ~0.1 mm-thick sterile polypropylene film was placed upon the inoculum on the tile to prevent dehydration and optimize bacterial contact with the tile. Following incubation at 37 °C for 24 h, 10 mL of soybean casein digest broth, containing lecithin and polyoxyethylene sorbitan monooleate (SCDLP), was used to collect the bacteria from each tile; colony-forming units (CFU) were enumerated after spreading serial dilutions of the initial suspension on Luria Bertani (LB) agar (15 g/L) medium. Each test was conducted in triplicate. The antibacterial activity (R-value) was calculated using the following equation: R = (Ut − U0) − (At − U0) = Ut − At, where U0 and Ut are the averages of the log (i.e., base 10 logarithm) of the number of CFU/cm^2 recovered from the tiles without silver, immediately after inoculation and after 24 h, respectively; At is the average of the log of the number of CFU/cm^2 recovered from the tiles with silver after 24 h. Each test was considered valid when the following equation was satisfied: $(L_{max} - L_{min})/(L_{mean}) \leq 0.2$ where L_{max}, L_{min}, and L_{mean} are the log of the maximum, minimum, and mean number of CFU found on the tile without silver immediately after inoculation, respectively.

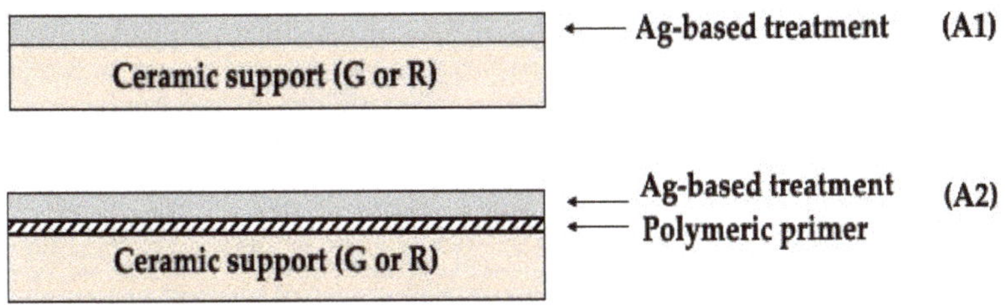

Figure 1. Schema of the investigated antibacterial systems.

Evaluation of stain resistance was carried out on all the samples by the method described in EN ISO 10545-14:2015 [37]. Staining agents (chrome green paste, alcoholic iodine solution (13 g/L), and olive oil) and cleaning procedures were applied to classify the samples' surface on the basis of detectable visual changes. Five levels of classification (from 5 to 1, respectively, the highest- and the lowest-performing) were used, as required in EN ISO 10545-14:2015.

The "sessile drop" technique was used to measure the wettability parameter of all the surface samples, using a multi-apparatus instrument, according to EN 15802:2010 [38]. Five drops, 2 μL in volume, of distilled water (wetting liquid) were uniformly deposited on the surfaces of the samples and singly captured. The average value of the contact angle was then derived, computing the uncertainty as the standard deviation for each samples' surface. By means of appropriate image-processing software, the contact angle was geometrically obtained by fitting the droplet's circular profile with a mathematical algorithm and calculating the angular coefficient of the tangent passing through the triple point (liquid-surface-air tensions).

Morphological and textural characteristics of the ceramic surfaces were evaluated using an optical profiler (LEICA DCM 3D, Heerbrugg, Switzerland), set on a metrological system (point autofocus profiling). The objective was set to 10× magnification to acquire, for each sample, an area of 4.66 mm × 4.36 mm with a 452 μm vertical scanning interval. Representative 3D elaborations were obtained, and numerical data were extrapolated in terms of peak density (Spd), according to ISO 25178-2:2012 [39]. In addition, 2D profiles were extrapolated from the acquired areas to obtain useful parameters such as average surface roughness (Ra), maximum peak height (Rp), and maximum valley depth (Rv), according to ISO 4287:2009 [40]. Microstructural and compositional surface information was obtained using scanning electron microscopy (SEM). Imaging was carried out using a field emission gun (FEG) instrument (Tescan, Mira3, Brno, Czech Republic), applying a working distance of 10 mm and a voltage of 10 kV. Microanalysis was performed using a scanning electron microscope (ZEISS EVO 40, Oberkochen, Germany) coupled with an energy dispersion spectroscopy system (EDS, Inca, Oxford, UK). All the samples were gold-coated before observation to make them conductive.

3. Results and Discussion

All the ceramic tiles tested in this work showed antibacterial activity against *E. coli* ATCC8739 (Table 3). In particular, G-A1, G-A2, and R-A2 showed strong antibacterial activity against this strain (bacterial reduction of 99.999% and R-values > 5.6), while R-A1 showed a lower effect, with an antibacterial activity of 99.71% and R-value > 2.5. The R-value of 5.88 corresponded to the absence of viable cells countable after the treatment.

Table 3. Antibacterial activities of the tiles tested in this work against *E. coli* ATCC8739.

	G-A1	G-A2	R-A1	R-A2
Bacterial reduction	99.999%	99.999%	99.71%	99.999%
R-value	5.880 ± 0.084	5.880 ± 0.084	2.538 ± 0.140	5.647 ± 0.008

These results indicate that the lack of a priming polymeric layer in A1 lowers the antimicrobial activity of porcelain stoneware tiles containing a high percentage of recycled material (R). Thus, although the antimicrobial activity of the R-type tiles was significant (R > 2.5), the presence of a polymeric primer seemed to further improve the final antimicrobial activity. Conversely, the absence of a polymeric base did not affect the antimicrobial efficacy of tiles made with typical raw materials (G).

Table 4 shows the stain-resistance classification of the investigated samples according to ISO 10545-14. Samples without antibacterial treatments were also tested as reference samples, with three staining agents (e.g., chrome green paste, alcoholic iodine solution, and olive oil, which were used to feature three different types of stains: powder, oxidizing, and film, respectively). All surfaces exhibited the maximal level of resistance to the oxidizing staining action, falling within class 5 for the iodine solution. Both G-A2 and R-A2 samples, to which antibacterial treatment was applied with a primer, exhibited the highest classification for all the staining agents, with results equal to or better than A1 and untreated samples.

Table 4. Stain resistance classes defined for the test samples according to ISO 10545-14.

Sample	CHROME *	IODINE	OIL
G	4	5	4
G-A1	4	5	4
G-A2	5	5	5
R	4	5	4
R-A1	4	5	3
R-A2	5	5	4

* The test is deemed to have also been passed if an oily halo remains, providing there is no longer visible green pigmentation.

The lowest class observed for stain resistance (class 3) was only attributed to the R-A1 sample when oil removal was tested, for which a deeper cleaning action was necessary to completely remove a light halo left by the film-staining agent.

Table 5 shows the average values of the contact angle measured for each sample. Overall, considering the standard deviation, these values fell within the range of 55–73°, which corresponds to an intermediate wettability degree for all of the surfaces under analysis. In general, this wetting behavior is typical for porcelain stoneware tiles [41].

Table 5. Average contact angle values and respective standard deviation for the analyzed samples.

	G	G-A1	G-A2	R	R-A1	R-A2
Contact angle (°)	60 ± 5	63 ± 7	62 ± 6	69 ± 3	65 ± 7	69 ± 4

Figure 2 shows the 3D topography scale models of the scanned areas of the analyzed surfaces. By comparing the reported topographies, it was noticeable that all surfaces were characterized by a heterogeneous structure due to a high absolute value of the gap between peaks and pits. This texture is in line with the productive process of unglazed ceramic tiles, which have a prominent micro-roughness [41].

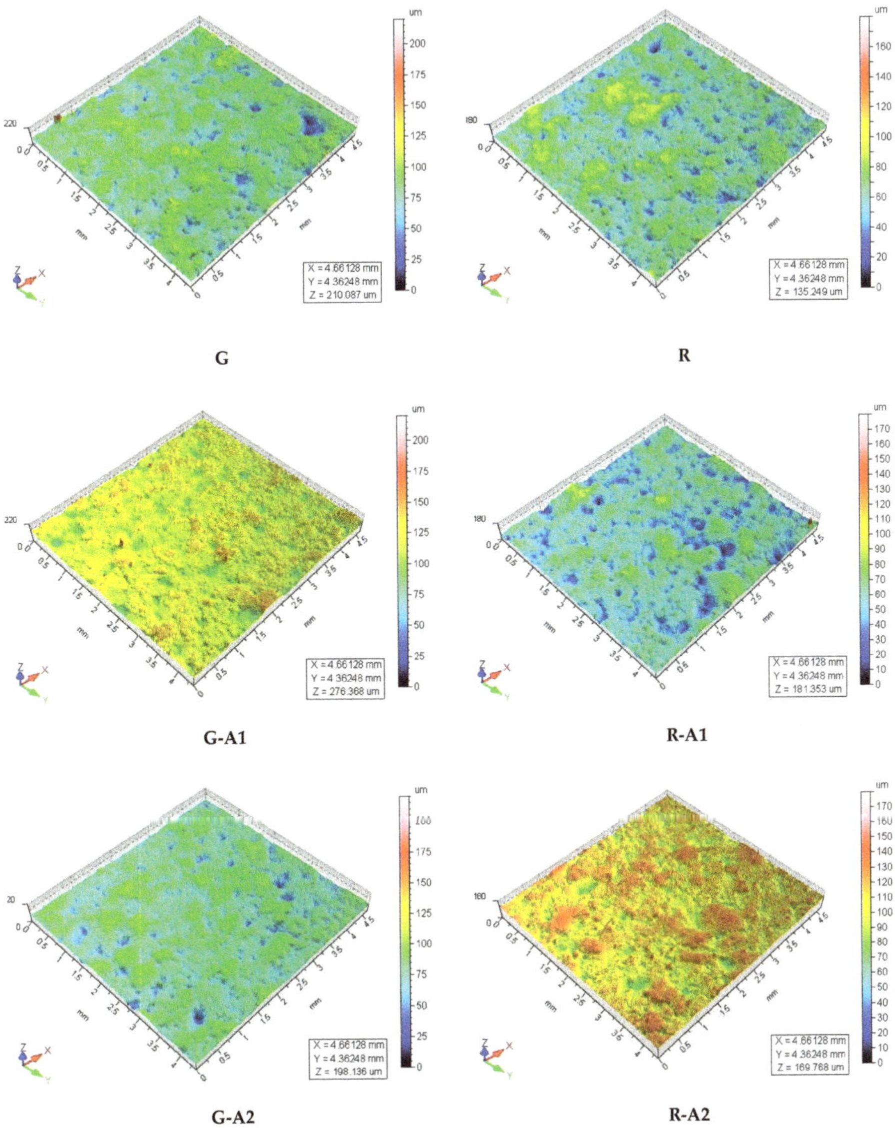

Figure 2. Graphic 3D elaborations of the investigated portion of the samples' surface.

The roughness parameters values, reported in Table 6, were not significantly different between untreated and treated samples. Moreover, in both G and R samples, the peak density, Spd, was slightly lower in treated samples than in untreated ones, with lower values in the A1-treated type, which was functionalized with only the silver treatment, i.e., without a primer layer. This aspect, which did not influence the antibacterial efficacy of the tested treatment, indicates that the silver-based treatment alone allowed a slightly smoother and more homogenous ceramic surface. Overall, the values of this parameter were higher for the R samples as compared to the G samples.

Table 6. Roughness parameters (Ra, Rv, Rp, and Spd) obtained by profilometric analysis on all the samples.

Parameter	Description	G	G-A1	GA2	R	R-A1	R-A2
Ra (μm)	Average roughness	4.8 ± 0.7	4.3 ± 0.6	4.6 ± 0.8	4.3 ± 0.8	4.4 ± 0.8	4.3 ± 0.6
Rv (μm)	Maximum valley depht	13 ± 3	12 ± 3	14 ± 4	12 ± 3	12 ± 3	12 ± 2
Rp (μm)	Maximum peak height	11 ± 2	10 ± 3	10 ± 2	10 ± 2	10 ± 2	12 ± 2
Spd (1/mm^2)	Density of peaks	9.0	4.8	7.5	17.6	8.4	10.0

Comparing the contact angle with the roughness values, some correspondence was appreciable. Similar to the average roughness (Ra), the contact angle and surface roughness values (except for the density of peaks, Spd) were comparable, also considering the standard deviation, in untreated and treated samples. These results indicate that silver-based antibacterial treatment did not influence the surface characteristics of the porcelain stoneware tile. Moreover, the scattering present in the results of both contact angle and average roughness was attributable to the heterogeneous nature of ceramic surfaces, which could cause significant spotted variations in measurements.

Even if it was not evident from the reported numerical roughness parameters, the 3D image of the R-A1 sample (Figure 2) seemed to be characterized by larger and deeper valleys compared to the other samples' images. This feature could be responsible for the worse antibacterial efficiency and stain resistance of this sample to oil, as reported in Tables 3 and 4, respectively. Indeed, grooved surfaces provide a favorable interface for bacterial colonization and the formation of biofilm. Biofilms are clusters of bacteria that are attached to a surface and are embedded in a self-produced matrix that provides protection against antimicrobial compounds and other environmental stresses [42].

The SEM images in Figures 3–5 show representative surfaces of the investigated samples at different magnifications. They revealed crystalline structures submerged in a vitreous matrix (flatter and smoothest portions), and, overall, a heterogeneous microstructure was noted in all the observed samples, confirming the previous surface features measured by profilometry (Table 6). This microstructure is typical of unglazed porcelain stoneware surfaces [43]. Detrital materials incoherent with the matrix were often observed, which could probably be related to particle fouling from the manufacturing process. Samples based on the G mix exhibited smoother surfaces compared to the relevant samples based on the R mix, except for samples G-A2 and R-A2, which looked very similar.

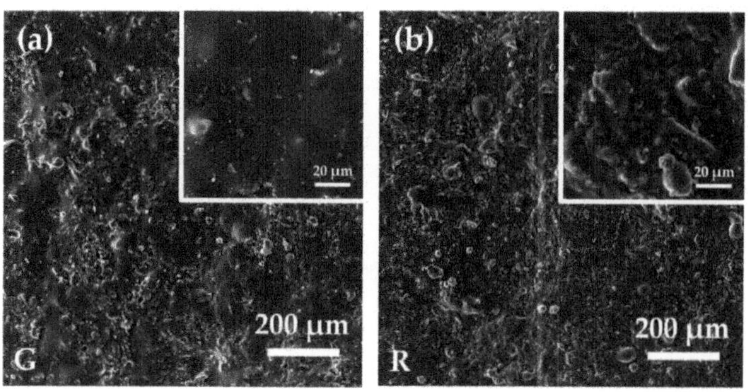

Figure 3. SEM images of G (**a**) and R (**b**).

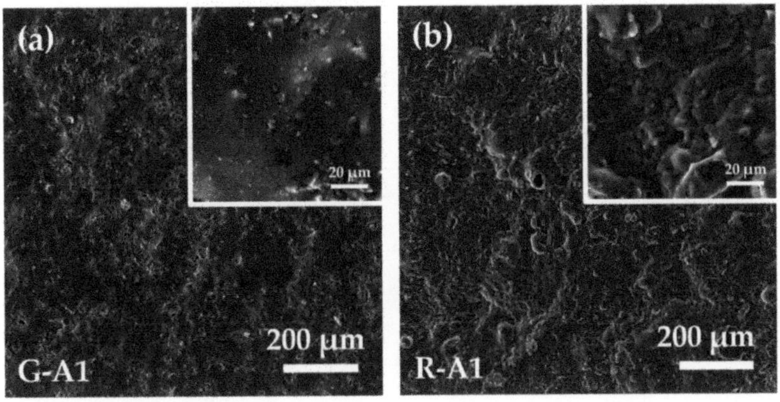

Figure 4. SEM images of G-A1 (**a**) and R-A1 (**b**).

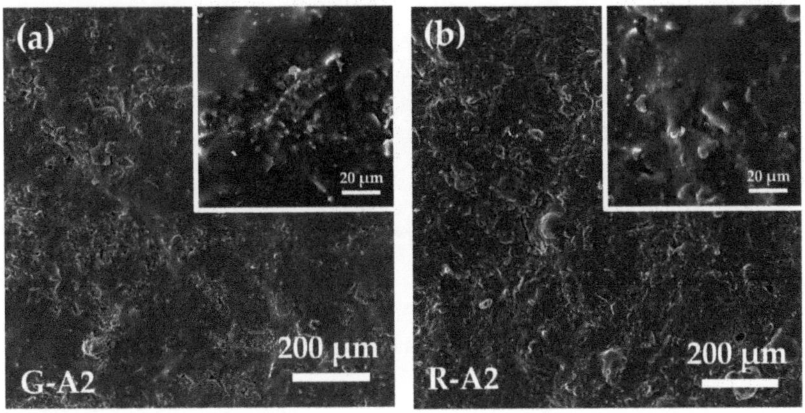

Figure 5. SEM images of G-A2 (**a**) and R-A2 (**b**).

Lastly, by observing the EDS spectra of all the surfaces in Figure 6, it was evident that the main compositional difference was the presence of Zircon (Zr) in G samples respective to R samples. This was related to the different compositions of raw materials used in G and R mixes. Indeed, the concentration of the detected elements (and therefore peak height) confirmed the weight percentages reported in Table 1. Conversely, silver was not experimentally or analytically detectable due to the low concentration applied (about 90 g/m^2) and the limited sensitivity of EDS analysis. In particular, by broadening the spectra, the acquisition software suggests that, if traceable, silver would give peaks for values of 2.98 and 3.15 keV, as shown in Figure 7.

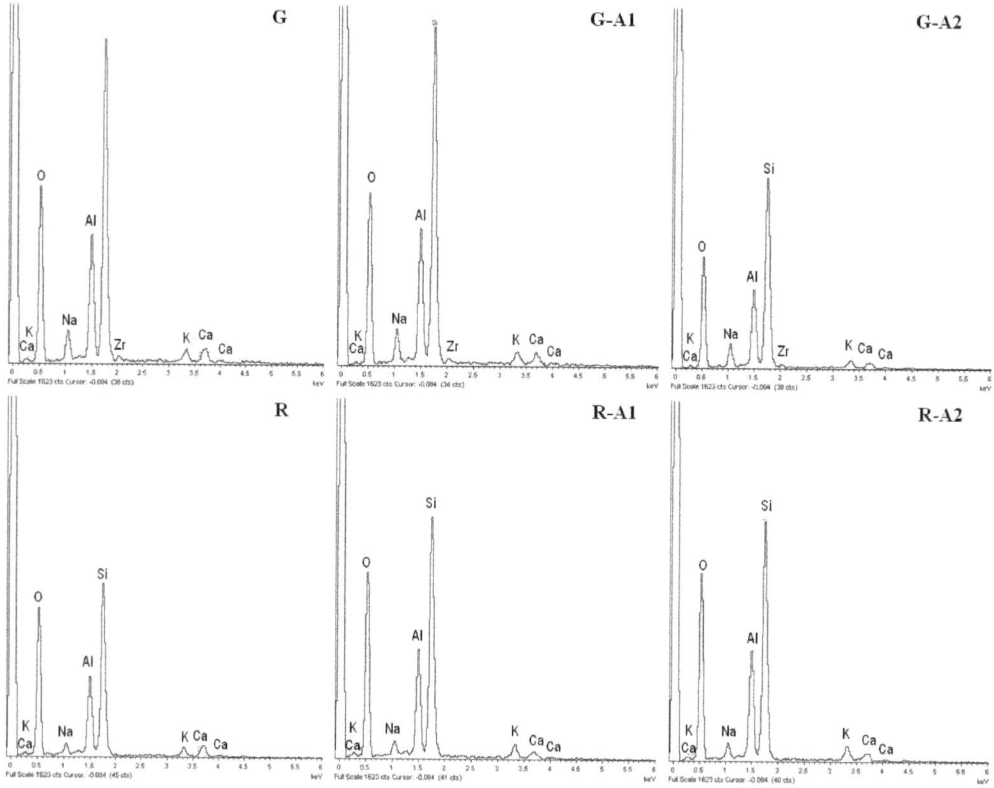

Figure 6. EDS spectra of treated samples G, G-A1, G-A2 and R, R-A1, R-A2.

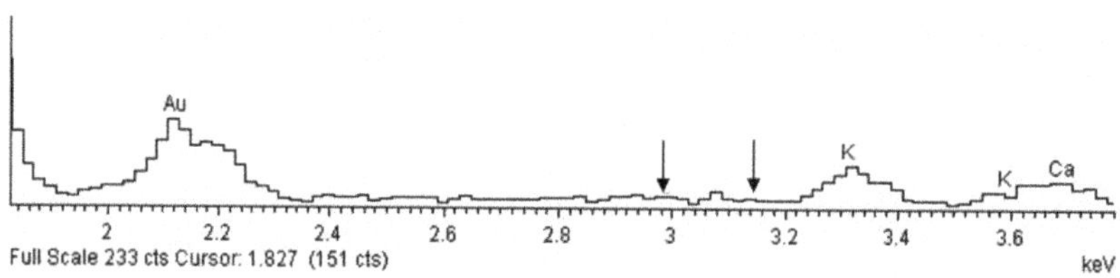

Figure 7. Enlarged EDS spectrum of the treated sample G-A1. The arrows indicate the keV typical of silver.

4. Conclusions

The results of this research allowed the evaluation of sustainable ceramic tiles with a high content of recycled material (R) in terms of high antibacterial efficacy, compared to traditional tiles without any waste material (G).

The silver-based functionalizing treatment (A1), applied with a screen-printing technique, proved to be effective for both types of ceramic substrate, even if a bit less for the R-A1 (99.71%) sample compared to G-A1 (99.999%). The addition of a polymeric primer (A2) improves performance in terms of microstructure and texture of the ceramic surface after firing. In particular, the microstructural and textural analyses showed that the cleanability (stain resistance) and antibacterial properties of porcelain stoneware tiles depend on surface characteristics such as roughness parameters.

In the specific case of the R-A1 sample, for which the antibacterial efficacy and stain resistance are the lowest, profilometry analysis, returned by 3D processing, highlights a greater number of valleys characterized by a slightly larger width than the surfaces of the other samples, and textural parameters showed the lowest Spd value (density of peaks) among the samples of the R series. Indeed, grooved surfaces offer an optimal substrate for bacterial adhesion and protection due to biofilm formation.

In conclusion, the correlation among data obtained from different experimental techniques highlights the importance of a multi-analytical approach for the definition of the surface performance of ceramic tiles. Moreover, our results showed that, independently of the body mix composition (with or without recycled raw materials), porcelain stoneware tiles can achieve strong antibacterial properties, indicating the possibility of the production of sustainable ceramics with a broad application range.

Author Contributions: Conceptualization, E.R.; methodology, V.L.T., E.R., G.M., S.N., D.G. and M.C..; formal analysis, D.G.; investigation, V.L.T. and S.N.; data curation, V.L.T.; writing—original draft preparation, V.L.T., G.M., E.R., S.N., D.G. and M.C.; writing—review and editing, E.R., G.M., M.C., D.G. and M.C.B.; supervision, for the experimental work E.R. and M.C.B., for the microbiological analysis M.C. All authors have read and agreed to the published version of the manuscript.

Funding: This research received no external funding.

Institutional Review Board Statement: Not applicable.

Informed Consent Statement: Not applicable.

Data Availability Statement: Not applicable.

Conflicts of Interest: The authors declare no conflict of interest.

References

1. Sciancalepore, C.; Manfredini, T.; Bondioli, F. Antibacterial and self-cleaning coatings for silicate ceramics: A review. *Adv. Nat. Sci. Nanosci. Nanotechnol.* **2019**, *92*, 90–99.
2. Seabra, M.P.; Grave, L.; Oliveira, C.; Alves, A.; Correia, A.; Labrincha, J.A. Porcelain stoneware tiles with antimicrobial action. *Ceram. Int.* **2014**, *40*, 6063–6070. [CrossRef]
3. Berto, A.M. Ceramic tiles: Above and beyond traditional applications. *J. Eur. Ceram. Soc.* **2007**, *27*, 1607–1613. [CrossRef]
4. Welch, K.; Sloan, G.; Ong, I.; Microban International. Antimicrobial Treatment for Ceramics and Sanitary Ware. *World Ceram. Rev.* **2017**, *121*, 101 108.
5. Bianchi, C.L.; Cerrato, G.; Bresolin, B.M.; Djellabi, R.; Rtimi, S. Digitally printed AgNPs doped TiO_2 on commercial porcelain-grès tiles: Synergistic effects and continuous photocatalytic antibacterial activity. *Surfaces* **2020**, *3*, 11–25. [CrossRef]
6. Baheiraeia, N.; Moztarzadeha, F.; Hedayati, M. Antibacterial Ag/SiO_2 thin film on glazed ceramic tiles prepared by sol-gel method. In Proceedings of the 18th Iranian Conference of Biomedical Engineering (ICBME), Tehran, Iran, 14–16 December 2011; IEEE: Piscataway, NJ, USA, 2011; pp. 105–108.
7. Bondioli, F.; Manfredini, T. Le nanopolveri nella funzionalizzazione delle superfici ceramiche. *Ceram. Inf.* **2008**, *468*, 231–237.
8. Slavin, Y.N.; Asnis, J.; Häfeli, U.O.; Bach, H. Metal nanoparticles: Understanding the mechanisms behind antibacterial activity. *J. Nanobiotechnology* **2017**, *15*, 65. [CrossRef]
9. Raghunath, A.; Perumal, E. Metal oxide nanoparticles as antimicrobial agents: A promise for the future. *Int. J. Antimicrob. Agents* **2017**, *49*, 137–152. [CrossRef]
10. Martinez-Gutierrez, F.; Olive, P.L.; Banuelos, A.; Orrantia, E.; Nino, N.; Sanchez, E.M.; Ruiz, F.; Bach, H.; Av-Gay, Y. Synthesis, characterization, and evaluation of antimicrobial and cytotoxic effect of silver and titanium nanoparticles. *Nanomedicine* **2010**, *6*, 681–688. [CrossRef]
11. Simon-Deckers, A.; Loo, S.; Mayne-L'hermite, M.; Herlin-Boime, N.; Menguy, N.; Reynaud, C.; Gouget, B.; Carriere, M. Size-, composition- and shape-dependent toxicological impact of metal oxide nanoparticles and carbon nanotubes toward bacteria. *Environ. Sci. Technol.* **2009**, *43*, 8423–8429. [CrossRef]
12. Karakoti, A.S.; Hench, L.L.; Seal, S. The potential toxicity of nanomaterials—The role of surfaces. *JOM* **2006**, *58*, 77–82. [CrossRef]
13. Choi, O.; Hu, Z. Size dependent and reactive oxygen species related nanosilver toxicity to nitrifying bacteria. *Environ. Sci. Technol.* **2008**, *42*, 4583–4588. [CrossRef] [PubMed]
14. Sohm, B.; Immel, F.; Bauda, P.; Pagnout, C. Insight into the primary mode of action of TiO_2 nanoparticles on Escherichia coli in the dark. *Proteomics* **2015**, *15*, 98–113. [CrossRef] [PubMed]
15. Fujishima, A.; Zhang, X.; Tryck, D.A. TiO_2 photocatalysis and related surface phenomena. *Surf. Sci. Rep.* **2008**, *63*, 515–582. [CrossRef]

16. Franzoni, E.; Bignozzi, M.C.; Rambaldi, E. TiO_2 in the building sector. In *Titanium Dioxide (TiO_2) and Its Applications*; Parrino, F., Palmisano, L., Dean, M., Eds.; Elsevier: Amsterdam, The Netherlands, 2021; pp. 449–479.
17. Chouirfa, H.; Bouloussa, H.; Migonney, V.; Falentin-Daudré, C. Review of titanium surface modification techniques and coating for antibacterial applications. *Acta Biomater.* **2019**, *83*, 37–54. [CrossRef]
18. Barmeh, A.; Nilforoushan, M.R.; Otroj, S. Wetting and photocatalytic properties of Ni-doped TiO_2 coating on glazed ceramic tile under visible light. *Thin Solid Film.* **2018**, *666*, 137–142. [CrossRef]
19. Da Silva, A.L.; Dondi, M.; Raimondo, M.; Hotza, D. Photocatalytic ceramic tiles: Challenges and technological solutions. *J. Eu Ceram. Soc.* **2018**, *38*, 1002–1017. [CrossRef]
20. Durán, N.; Durán, M.; de Jesus, M.B.; Seabra, A.B.; Fávaro, W.J.; Nakazato, G. Silver nanoparticles: A new view on mechanisti aspects on antimicrobial activity. *Nanomedicine* **2016**, *12*, 789–799. [CrossRef]
21. Epple, M.; Chernousova, S. Silver as Antibacterial Agent: Ion, Nanoparticle, and Metal. *Angew. Chem. Int.* **2013**, *52*, 1636–1653.
22. Jung, W.K.; Koo, H.C.; Kim, K.W.; Shin, S.; Kim, S.H.; Park, Y.H. Antibacterial Activity and Mechanism of Action of the Silver Io in *Staphylococcus aureus* and *Escherichia coli*. *Appl. Environ. Microbiol.* **2008**, *74*, 2171–2178. [CrossRef]
23. Sim, W.; Barnard, R.T.; Blaskovich, M.A.T.; Ziora, Z.M. Antimicrobial Silver in Medicinal and Consumer Applications: A Paten Review of the Past Decade (2007–2017). *Antibiotics* **2018**, *7*, 93. [CrossRef]
24. Sheng, H.; Congqin, N.; Yue, Z.; Lei, C.; Kaili, L.; Jiang, C. Antibacterial Activity of Silicate Bioceramics. *J. Wuhan Uni Technol.-Mat. Sci. Edit.* **2011**, *26*, 226–230.
25. Kummala, R.; Brobbey, K.J.; Haapanen, J.; Mäkelä, J.M.; Gunell, M.; Eerola, E.; Huovinen, P.; Toivakka, M.; Saarinen, J.J. Antibacteria activity of silver and titania nanoparticles on glass surfaces. *Adv. Nat. Sci. Nanosci. Nanotechnol.* **2019**, *10*, 015012. [CrossRef]
26. Petronella, F.; Truppi, A.; Striccioli, M.; Curri, M.L.; Comparelli, R. Photocatalytic Application of Ag/TiO_2 Hybrid Nanoparticle In *Noble Metal-Metal Oxide Hybrid Nanoparticles, Fundamentals and Applications; Satyabrata Mohapatra, Phuong Nguyen-Tri, Tuan An Nguyen*; Elsevier: Amsterdam, The Netherlands, 2019; pp. 373–394.
27. Durango-Giraldo, G.; Cardona, A.; Zapata, J.F.; Santa, J.F.; Buitrago-Sierra, R. Titanium dioxide modified with silver by tw methods for bactericidal applications. *Helyon* **2019**, *5*, e01608. [CrossRef]
28. Djellabi, R.; Basilico, N.; Delbue, S.; D'Alessandro, S.; Parapini, S.; Cerrato, G.; Laurenti, E.; Falletta, E.; Bianchi, C.L. Oxidativ inactivation of SARS-CoV-2 on Photoactive AgNPs@TiO_2 Ceramic Tiles. *Int. J. Mol. Sci.* **2021**, *22*, 8836. [CrossRef]
29. Confindustria Ceramica. *Indagini Statistiche Sull'industria Italiana*; Confindustria Ceramica: Sassuolo, Italy, 2020; pp. 8–35.
30. Palmonari, C. *Il Grès Porcellanato*; Centro Ceramico: Bologna, Italy, 1989.
31. Zannini, P. Treatments for ceramic surfaces. *Ceram. World Rev.* **2014**, *107*, 62–64.
32. Dondi, M.; Garcia-Ten, J.; Rambaldi, E.; Zanelli, C.; Vicent-Cabedo, M. Resource efficiency versus market trends in the cerami tile industry: Effect on the supply chain in Italy and Spain. *Resour. Conserv. Recycl.* **2021**, *168*, 105271. [CrossRef]
33. Rambaldi, E.; Esposito, L.; Tucci, A.; Timellini, G. Recycling of polishing porcelain stoneware residues in ceramic tiles. *J. Eu Ceram. Soc.* **2007**, *27*, 3509–3515. [CrossRef]
34. Rambaldi, E.; Fazio, S.; Prete, F.; Bignozzi, M.C. Innovative ceramic tile mixes. *Cfi/Ber. DKG* **2016**, *93*, 57–60.
35. Rambaldi, E. Pathway towards a high recycling content in traditional ceramics. *Ceramics* **2021**, *4*, 486–501. [CrossRef]
36. International Organization for Standardization, ISO 22196: Measurement of antibacterial activity on plastics and other non-porou surfaces. International Organization for Standardization: Geneva, Switzerland. 2011. Available online: https://www.iso.org standard/54431.html (accessed on 20 July 2020).
37. International Organization for Standardization, ISO10545-14: Ceramic tiles-Part14: Determination of resistance to stains.International Organization for Standardization: Geneva, Switzerland. 2015. Available online: https://www.isc org/standard/60972.html (accessed on 20 June 2021).
38. Comité Euroéen de Notmalisation, EN 15802: Conservation of cultural property-test methods-Determinaation of static contac angle. Comité Euroéen de Notmalisation: Bruxelles, Belgium. 2010. Available online: https://standards.iteh.ai/catalog standards/sist/c3a0a8e7-5437-499a-89e2-c4355aae5534/sist-en-15802-2010 (accessed on 20 June 2021).
39. International Organizatio for Standardization, ISO 25178-2: Geometrical Product Specifications (GPS)—Surface Texture: Areal—Part 2: Terms, Definitions and Surface Texture Parameters. International Organization for Standardization: Geneva, Switzerland 2012. Available online: https://www.iso.org/standard/42785.html (accessed on 20 March 2021).
40. Intenational Organization for Standardization, ISO 4287: Geometrical Product Specifications (GPS)—Surface Texture Profile Method—Terms, Definitions and Surface Texture Parameters; International Organization for Standardization Geneva, Switzerland. 2009. Available online: https://www.iso.org/standard/10132.html (accessed on 20 March 2021).
41. Rambaldi, E.; Lucchese, B.; Engels, M.; Bignozzi, M.C. Evaluation of durability and cleanability performances of protective treatments for lapped ceramic tiles—Part 2. *Int. J. Appl. Ceram. Tech.* **2018**, *16*, 625–637. [CrossRef]
42. Donlan, R.M.; Costerton, J.W. Biofilm: Survival mechanisms of clinically relevant microorganisms. *Clin. Microbiol. Rev* **2002**, *15*, 167–193. [CrossRef] [PubMed]
43. Dondi, M.; Ercolani, G.; Guarini, G.; Melandri, C.; Raimondo, M.; Rocha e Almedra, E.; Tenorio Cavalcante, P.M. The role of surface microstructure on the resistance to stains of porcelain stoneware tiles. *J. Eur. Ceram. Soc.* **2005**, *24*, 357–365. [CrossRef]

coatings

MDPI

Article

New Strategy for Creating TiO_2 Thin Films with Embedded Au Nanoparticles

Sofia Rubtsov [1], Albina Musin [2], Michael Zinigrad [1], Alexander Kalashnikov [3] and Viktor Danchuk [1,*]

1 Department of Chemical Engineering, Biotechnology and Materials, Faculty of Engineering, Ariel University, Ariel 4076414, Israel; Sofiar@ariel.ac.il (S.R.); zinigradm@ariel.ac.il (M.Z.)
2 Physics Department, Faculty of Natural Sciences, Ariel University, Ariel 4076414, Israel; albinam@ariel.ac.il
3 Rock Capital Partners, 101000 Moscow, Russia; alexander.kalashnikov@gmail.com
* Correspondence: viktorde@ariel.ac.il

Abstract: This paper proposes a new strategy for producing thin films of TiO_2 with embedded gold nanoparticles (TiO_2/AuNP). One of the main tasks was the synthesis of a stable dispersion of TiO_2 and gold nanoparticles in an aqueous solution of ethylene glycol, suitable for inkjet printing—ink with complex gold nanoparticles (AuCNP ink). The AuCNP were synthesized by a reduction from tetrachloroauric acid in the presence of TiO_2 nanoparticles and ethylene glycol (EG). The final formation of TiO_2/AuNP films occurred during the annealing of AuCNP layers, inkjet printed on a glass substrate. The TiO_2/AuNP films demonstrate absorbance in the yellow-green range due to the localized surface plasmon resonance (LSPR) and are promising for solar cell application.

Keywords: gold nanoparticles; inkjet printing; localized surface plasmon resonance; perovskite solar cells

1. Introduction

Gold nanoparticles (AuNPs) have a large field of potential applications due to their unique optical and electrical properties, different from their bulk form in many ways. AuNPs have abundant applications in various branches of science, including medicine, materials science, biology, chemistry, and physics [1,2]. Thin-film solar cells with a total thickness of 1–2 μm are of great interest because of the variety of available methods for manufacturing thin layers, the possibility of using different types of materials and substrates, and the economy of materials. The weak point of all thin-film solar cells is that they do not absorb as much light as the massive layer. Therefore, it is necessary to develop methods for capturing light in the submicron layer to enhance the photocurrent and improve the efficiency of the solar cell. One of the methods of light capture is based on the introduction of metallic nanostructures in which LSPR can be excited. The metal nanoparticles (NPs) scatter light, increasing the effective path length within a layer and enhancing light absorption. In addition, hot electrons, induced by plasmons, can facilitate electron transport entering the conduction band of electron transport material [3–5]. Au and Ag metallic nanostructures are the most commonly used because they have good chemical stability and plasmon resonance frequencies in the visible range [6]. It was demonstrated that AuNPs improve device performance in thin-film solar cells of perovskite-type, organic, dye-sensitized ones [6–9]. Therefore, the method of deposition of thin films of metal oxide with embedded AuNPs, proposed in this work, is promising for thin-film solar cells applications.

AuNPs of the desired size, shape and spatial arrangement are required for plasmonic applications. Methods of deposition of AuNPs include spin-coating of NPs synthesized with various chemical routes [3,6,10], sputtering [11,12], and lithographic techniques [13,14]. A promising method for introducing plasmonic structures into the architecture of an electric generating cell is inkjet printing [15]. This method allows one to control the concentration of NPs in the layer. The introduction of NPs may be performed simultaneously with the deposition of a dielectric matrix, which serves as a functional layer. AuNPs can be prepared by several bottom-up liquid phase routes. One of the most popular methods is the chemical reduction of the precursor salts by sodium citrate (Turkevich's method [16,17]) or other reducing agents, e.g., sodium borohydride [18,19]. After the completion of a chemical reaction, contaminant inorganic salts and ions remain in the product and can be removed only by precipitation and washing of synthesized nanoparticles. The need for multiple purifications of the obtained NPs to ensure a required purity is a drawback for further use in functional layers. Along with the citrate method, the so-called polyol process was developed when polyalcohols were proposed as a class of compounds to produce metal NPs, suitable for various applications such as inks and functional protective coatings. In this process, polyol refers to a diol, such as EG and its derivatives [20]. Basically, a polyol synthesis reduces a metal salt precursor by a polyol species and has certain advantages. Due to the high boiling point of the reducing medium, the synthesis can be carried out at a relatively high temperature. Metal particles are protected from oxidation when they remain in the medium. The high viscosity of the medium and its ability to coordinate metal precursors can minimize coalescence and promote a diffusion-controlled regime for particle growth. All these conditions favor the production of size-controlled crystalline particles. As a rule, the use of the polyol alone is not sufficient to control the size and shape of the particles, so capping agents are added to obtain a desired distribution of NPs. In the early works [21], spherical AuNPs were synthesized in EG in the presence of polyvivnylpyrrolidone (PVP), which prevented aggregation of NPs. It was shown that the size of the AuNPs might be controlled by varying the concentration of nonionic surfactant Tween 80 (polyethylene glycol sorbitan monooleate) during the reduction of tetrachloroauric acid by maltose [1]. Molecules of Tween 80 interacted with growing gold nuclei, formed during the initial phase of the reduction process, and blocked their surface. The formation of new nuclei was preferred over the further growth of the preformed ones.

An important class of materials, usually used as supported metal catalysts, includes the NPs deposited/embedded onto a solid support. The support material is essential in determining the catalytic behavior of supported AuNPs; both organic polymers and inorganic materials got quite a lot of attention in the literature [22,23]. In a typical liquid-phase synthesis procedure for metal oxide supported AuNPs, preformed metal oxide particles are mixed with an aqueous solution of $HAuCl_4$, and the necessary additives, stirring is performed, followed by washing a suspension, drying, and calcining. Similar methods have been used to deposit Au on oxide supports such as TiO_2, Fe_2O_3, Al_2O_3, MgO, and others [22,24,25]. Reduction of Au(III) to Au(0) presumably occurs in several stages. It begins with the adsorption of gold chloride and gold chloro-hydroxy $[Au(OH)_{4-x}Cl_x]^-$ complexes on a surface of metal oxide particles via electrostatic interaction or ligand exchange. It was suggested [25,26] that the gold chloro-hydroxy species interact with surface M–OH groups present at the oxide–water interface by the mechanism:

$$MOH + [Au(OH)_{4-x}Cl_x]^- \rightarrow MAu(OH)_{3-x}Cl_{x-1} + H_2O + Cl^- \quad (1)$$

where M is a metal, and the OH group is the result of the dissociative water chemisorption. With some kind of reducing agent or by irradiation, the Au complex can be subsequently reduced to Au clusters, agglomerating to AuNPs on the surface of the support oxide particles. It was demonstrated [26] that $HAuCl_4$ reduced spontaneously in EG in the presence of preformed ZnO nanoparticles at room temperature, without any additional stabilizers, forming Au/ZnO nanostructures with a pronounced plasmonic band in the absorption spectra. Considering the reaction (1), the authors [26] assumed that the electrostatic attraction between gold chloride (or chloro-hydroxyl complexes) and a positively charged ZnO surface promoted close contact and chemical adsorption of gold precursors by replacing one of the chloride ligands with a hydroxyl group on the ZnO surface. The reaction between Au and EG led to the reduction of Au(III) to Au(I) and eventually to Au(0). The further agglomeration of Au(0) species resulted in the formation of Au NPs, which was confirmed by the appearance of a plasmon band in the absorption spectrum of the solution of precursors. It was noticed that the plasmon band appeared only after the decay of the absorbance of Au(III) species [26]. In the liquid-phase reduction method, the combination of components and the order of their introduction into the reactor affects the size of Au NPs and, accordingly, the catalytic activity of the obtained materials [27]. Typically, small Au NPs (2–5 nm) have higher activity, and the catalytic performance is sensitive to the preparation method and residual chloride content. The metal oxides supported reduction is a recognized method for preparing catalysts, so the catalytic properties of Au/metal oxide complex NPs are the most studied [28,29].

However, similar techniques can be used to synthesize NPs for other applications, particularly for introducing them into thin layers by printing or spin coating to create functional layers with LSPR properties. In this work, we aimed to develop a method for synthesizing ink, that is, a stable dispersion of AuCNP in a liquid medium, to print thin films based on TiO_2 with embedded AuNPs. This ink may be further used for inkjet printing without any additional operations. After deposition and annealing, thin TiO_2 layers, including AuNPs surrounded with TiO_2 (TiO_2 with embedded gold nanoparticles, TiO_2/AuNP), may be manufactured. The interest in a method for synthesizing print-ready inks is twofold. First, it can provide a direct synthesis of Au NPs with a functionalized surface. Second, the kinetics of AuNPs growth should be determined by both the synthesis and deposition conditions, that is, a degree of surface coverage during the deposition of a layer and further heat treatment processes. We assume that this method of manufacturing functional layers can find a wide range of applications, for example, in thin-film solar cells, sensors, and molecular electronics.

2. Experimental Section

The TiO_2 NPs (TNP) were synthesized with a sol-gel method, described in the Supplementary Materials (SM), using titanium isopropoxide ($Ti[OCH(CH_3)_2]_4$, 99.995%, Alfa Aesar, Lancashire, UK), nitric acid (HNO_3 70%, Daejung, Siheung-si, Korea) and isopropanol (C_3H_8O Carlo Erba, Cornaredo, Italy). Ink with gold complex nanoparticles (AuCNP) was synthesized from hydrogen tetrachloroaurate(III) trihydrate ($HAuCl_4 \cdot 3H_2O$ 99.99%, Alfa Aesar), ethylene glycol (Daejung), and surfactant Tween 80 (Sigma-Aldrich, Hamburg, Germany). Bidistilled water (DI) (Milli-Q® IQ 7003, Merck, Darmstadt, Germany) was used in all experimental procedures.

2.1. Preparation of AuCNP Ink

The ink presenting a dispersion of AuCNP and TNP in the mixture of EG and water was prepared with a modified polyol method on TiO_2 support [25]. The 0.125 g of TiO_2 NPs xerogel (see Supplementary Materials) was dissolved in 6 mL EG and 2 mL DI water. The prepared colloid solution was heated till 150 °C in a flask immersed in an EG bath; vapors were cooled in a jacketed double coil condenser (44 cm length) with circulating water at 15 °C and returned to the flask. The acid solution consisted of 0.0677 g $HAuCl_4 \cdot 3H_2O$ mixed with 2 mL DI water and 1.2 μL Tween 80, which was quickly injected into the flask with boiling TiO_2 colloid. The mixture was boiled for 10 min under constant intensive stirring (500 rpm) with a magnetic stirrer. Next, the resulting dispersion of AuCNP was cooled to room temperature and stored at 8 °C in a lab refrigerator.

2.2. Printing Methods

Continuous AuCNP ink layers were deposited on 25 mm × 25 mm Soda Lime (Corning, Corning, NY, USA) glass substrates with Dimatix Materials printer DMP 2850 (Santa Clara, CA, USA) with a 10 pL cartridge. The substrates were preliminary cleaned according to the protocol, presented in Supplementary Materials, and treated for 2 min in RF plasma (Plasma Cleaner, MTI Corporation, Richmond, CA, USA) in residual oxygen at air pressure in the chamber at 200 mTorr, "High" mode for 2 min. Before filling a cartridge, the AuCNP ink was filtered with a 0.22 μm filter (Jet Biofil, Guangzhou, China). The printing process was carried out with a jetting frequency of 5 kHz, and the voltage on the nozzles was varied in the range of 19 to 25 V. The ink was printed on glass substrates with a drop spacing of 40 μm in three layers. As the contact diameter of a single drop on glass was about 80 μm, that is, larger than the drop spacing, jetted drops merged and formed a continuous layer. After deposition, the samples presenting a continuous thin layer of ink on glass substrate were sintered at 550 °C for 30 min with a continuous 10 sccm oxygen flow (ALIMC-500 ccm, Alicat Scientific, Richmond, CA, USA) in the tube furnace KJ-T1200 (Zhengzhou Kejia Furnace Co., Ltd., Zhengzhou, China). Estimates carried out on the assumption that all components of the ink had completely reacted and all AuNPs were embedded in the TiO_2 matrix after printing and annealing, give 23 wt% of the Au content in the TiO_2 matrix after film annealing.

2.3. Characterization Methods

Particle size and zeta potential values were measured using Lightsizer™ 500 instrument (Anton Paar GmbH, Graz, Austria). All dispersions were diluted 1:30 with DI water. Particle size measurements were performed by the dynamic light scattering (DLS) method at 10 °C, the refractive index of a solvent was set as for water [30]. The stability of NPs in dispersion was evaluated by the zeta potential using electrophoretic light scattering (ELS) [31]. The ELS measurements were made at 25 °C with a maximum voltage of 200 V. The rheological and physical properties were evaluated to determine the ink printability. The viscosity of the AuCNP ink was evaluated with a DVE viscometer (Ametek Brookfield, Middleborough, MA, USA) at 100 rpm of spindle S18. A Sigma Attension 703D tensiometer (Biolin Scientific, Gothenburg, Sweden) was used for the surface tension measurements by Du Nouy's method. The density of the AuCNP dispersion was calculated from mass and volume measurements with a pycnometer 10 mL. All measurements were completed at the same conditions at 25 °C.

Measurements of acidity/basicity of ink solutions with and without adding $HAuCl_4$—pH and oxidation rate reaction (ORP)—were carried out with a Multi parameter Analyzer DZB-712 (INESA Scientific Instrument, Shanghai, China) at room temperature.

The contact angle was measured with optical microscope MXB 5000REZ (Hirox Co., Tokyo, Japan), for a 5 μL droplet placed on a glass substrate, cleaned the same way as for printing. Measurements were made for three solutions: EG:H_2O 3:2 (liquid base of the ink), TiO_2 colloidal NPs in the liquid base, and AuCNP ink.

Thermal decomposition and compounds scans experiments were performed with a thermogravimetric analysis (TGA) (Pyris 1, PerkinElmer, Waltham, MA, USA). TGA measurements were carried out in the temperature range from 35 to 600 °C, at a heating rate of 10 °C/min, and a gas flow (N_2) of 20 mL/min.

The morphology of AuCNP ink films was studied with a TESCAN MAIA3 (Brno, Czech Republic) scanning electron microscope in SEM and scanning transmission electron microscopy (STEM) mode. The crystalline structure of the films was determined with an XRD Rigaku SmartLab SE (Rigaku, Tokyo, Japan) diffractometer with a CuKα source. The size distribution of AuNP embedded in the TiO_2 matrix was evaluated by analysis of SEM images with ImageJ (version V 1.53, 2021, University of Wisconsin, Madison, WI, USA) software using a modified "Particle size" plugin. The absorbance of the films was measured with a spectrophotometer UV-Vis GENESYS 10 (Thermo Scientific, Waltham, MA, USA) in the wavelength interval from 400 to 800 nm. The Raman spectra were recorded with Enspectr R532 (Enhanced Spectrometry Inc., Meridian, ID, USA) in a scanning range of 160–1600 cm^{-1}.

3. Results and Discussion

3.1. Characterization of the AuCNP Ink

The synthesized dispersion of AuCNP, including inorganic and organic species, was stable and repeatable. To study the composition of the dispersion, five types of solutions were prepared, described below, and their parameters were measured and compared:

- N1—The solution of EG/H_2O 3:2 by volume, that is the ink's base liquid components.
- N2—The colloid solution of TNP in EG/H_2O 3:2 without adding $HAuCl_4$.
- N3—The AuCNP precursor, which is the mixture of all ink components without heating and boiling, with stirring only.
- N4—The water AuCNP ink without EG.
- N5—The AuCNP ink.

The detailed description of the samples N1–N5 is presented in Supplementary Materials (see Figure S1).

The mean hydrodynamic diameter (HDD) of particles, measured with "Lightsizer™ 500", for N3, N4, and N5 dispersions, is presented in Figure 1A. Immediately after the synthesis, the HDD of AuCNP ink (N5) was about 270 nm with a relatively wide size distribution. Unlike the N5 ink, the scattering plots from the N4 and N3 dispersions include two maxima, indicating the presence of two groups of particles of different sizes in the solutions. The first group with an average diameter of 40 nm corresponds to TNP (see Figure S2), and the other (370 nm) to AuCNP. Therefore, adding EG and heating and boiling the ink precursor is necessary for forming the homogeneous AuCNP dispersion.

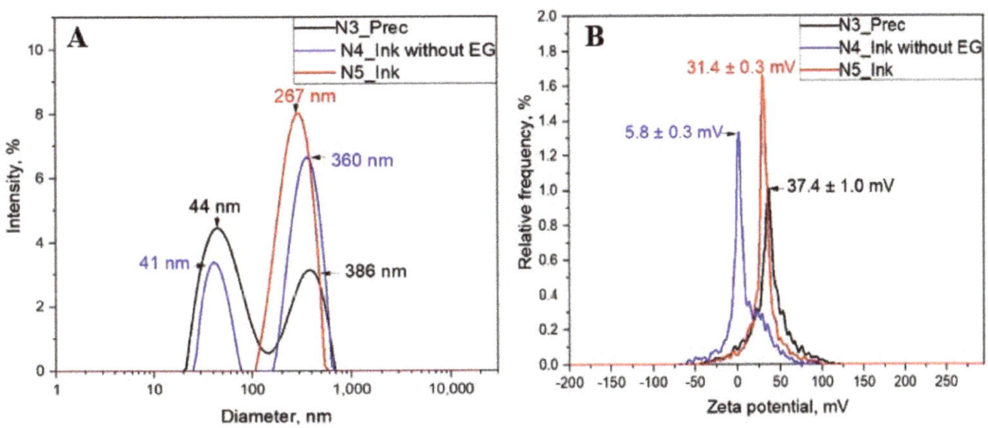

Figure 1. (**A**)—Particle size distribution and (**B**)—zeta potential of colloidal AuCNP obtained by DLS and ELS, respectively, ink precursor—black curves, AuCNP ink—red curves, and AuCNP ink without EG—blue curves.

The values of zeta potential and pH of the N2–N5 dispersions are presented in Figure 1B and Table 1, correspondingly. Let us pay attention to the fact that the pH for all these dispersions is acidic (see Table 1) since they are based on the N2 dispersion. The zeta potentials of the samples N2, N3, and N5 differ insignificantly and are close to 30 mV, which, according to [31], corresponds to dispersions with optimal stability, whereas the zeta potential of the N4 dispersion is much less and equals 5.8 mV, which is one of the reasons for its instability. Indeed, visible precipitation of the N4 dispersion occurred within an hour after preparation, while the N2 and N5 dispersions remained stable even after several months. This fact indicates the decisive role of EG in the formation of stable AuCNP ink. We supposed that the particles with a size range from 250 to 400 nm are an agglomerate of TiO_2 NPs, nascent Au clusters, ionic species of $HAuCl_4$ decomposition, and EG molecules (AuCNP) [25–27,32–35].

Table 1. pH and zeta potential values.

Sample	Composition	pH, ±0.01%	Zeta Potential, mV
N2	TiO_2 colloid	0.45	30.0 ± 0.3
N3	Ink precursor	1.04	37.4 ± 1.0
N4	AuCNP without EG	1.94	5.8 ± 0.3
N5	AuCNP ink	0.62	31.4 ± 0.3

The Raman spectra of the N1, N2, and N5 dispersions are presented in Figure 2. Recall here that the N5 is the final AuCNP ink, the N2 is the TNP (colloid solution of TiO_2 NPs in EG/H_2O 3:2), and the N1 is the solution of EG/H_2O 3:2 by volume, which is the liquid base for both N5 and N2. The Raman lines at 348, 481, and 525 cm^{-1} present in the spectra for all dispersions were assigned to a C–C–O torsional and bending modes in EG [36].

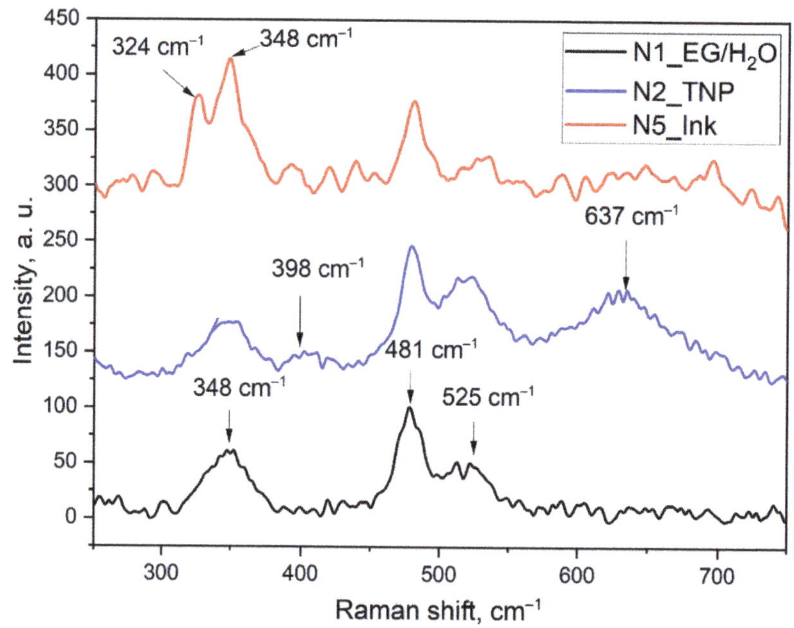

Figure 2. The Raman spectra of the N1, N2, and N5 dispersions.

Comparative analysis of the presented Raman spectra revealed two characteristic features. The first is the presence of bands at 324 and 348 cm^{-1} in the spectrum of N5, which correspond to the stretching of Au–Cl in $[AuCl_4]^-$, indicating the presence of these species in the AuCNP ink even after heating and boiling [32]. It should be emphasized here that the 348 cm^{-1} peak is also present in the Raman scattering spectrum of EG. However, for the N5 dispersion, this peak is more intense than for N1 and N2 since it is a superposition of the corresponding EG and Au–Cl peaks. Thus, this suggests that during the synthesis of the ink, some of the gold remains unreduced. In the N5 spectrum, the absence of peaks in the range of 560–580 cm^{-1}, related to the stretching vibrations of Au–OH, is consistent with the chloro-hydroxyl compounds of gold identified in solutions containing gold(III) chloride at low pH [32,33]. Note that the isoelectric point (IEP) of the support is of great importance in the metal-oxide-assisted reduction of Au NPs. When the pH of the solution is higher than its IEP, the adsorption of cations on the surface of the support particles occurs; otherwise, anions are adsorbed [34]. The IEP of TiO_2 is 6.0, and the pH values of both N5 and N2 dispersions are less than 1 (see Table 1); therefore, it is quite possible that the formation of a surface gold complex is initiated by the $[AuCl_4]^-$ anion adsorption on the TiO_2 surface. The other feature is the presence of peaks at 398 and 637 cm^{-1} in the N2 spectrum, which are characteristic of the TiO_2 anatase phase [35], and a significant reduction in the intensity of these peaks in the N5 dispersion. This fact is one more evidence of TNP participation in the formation of AuCNP agglomerates.

For ink, in addition to the characteristics of suspension, surface tension and viscosity are important parameters. Surface tension, viscosity, and density of the dispersion are the basic properties defining the formation and behavior of liquid jets and drops. The combination of these parameters, expressed by the dimensionless Ohnesorge number, is usually used for evaluating the printability of ink:

$$\mathrm{Oh} = \frac{1}{Z} = \frac{\eta}{\sqrt{\gamma \rho d}} \quad (2)$$

where η is the fluid's viscosity, γ—surface tension, ρ—density, and d—the characteristic dimension (diameter of the jet nozzle or drop). The rheological properties of the ink are presented in Table 2. The Z parameter for the reciprocal Ohnesorge number defines the range over which liquids can be printed as $10 > Z > 1$ [37–39]. d is the size of the silicon nozzle and is approximately 21.5 μm. It can be concluded (see Table 2) that the rheological properties of the AuCNP ink are in a stable range for printing.

Table 2. Rheology parameters of the synthesized AuCNP ink.

Sample	Composition	Viscosity, cP, ±0.2	Surface Tension, N/m, ±0.5	Density, g/cm^3, ±0.05	Z, ±0.1	Contact Angles, °, ±0.5
N1	EG/H_2O 3:2	4.9	54.2	1.10	7.30	13.9
N2	TNP	5.2	53.4	1.11	6.87	16.7
N5	AuCNP ink	9.7	52.8	1.11	3.60	18.3

Besides the rheological properties, the wettability is also critical for the formation of a uniform layer. The contact angle measurements presented in Table 2 show that the contact angles of the TNP and AuCNP inks are only slightly larger than for the EG/H_2O solution, and in all cases, are less than 20°. Consequently, TNP and AuCNP dispersions perfectly spread over the surface of the glass substrate, which contributes to the formation of uniform continuous layers.

In addition to the dispersion characteristics and the methods and modes of deposition of layers, the determining factor for forming homogeneous uniform films is the procedure of their temperature treatment (drying and annealing). In order to correctly set the parameters of the temperature treatment of the layers and reveal the process of the film formation, a TG analysis of dispersions N1, N3, and N5 was carried out. The TG and DTG data are presented in Figure 3A,B, respectively.

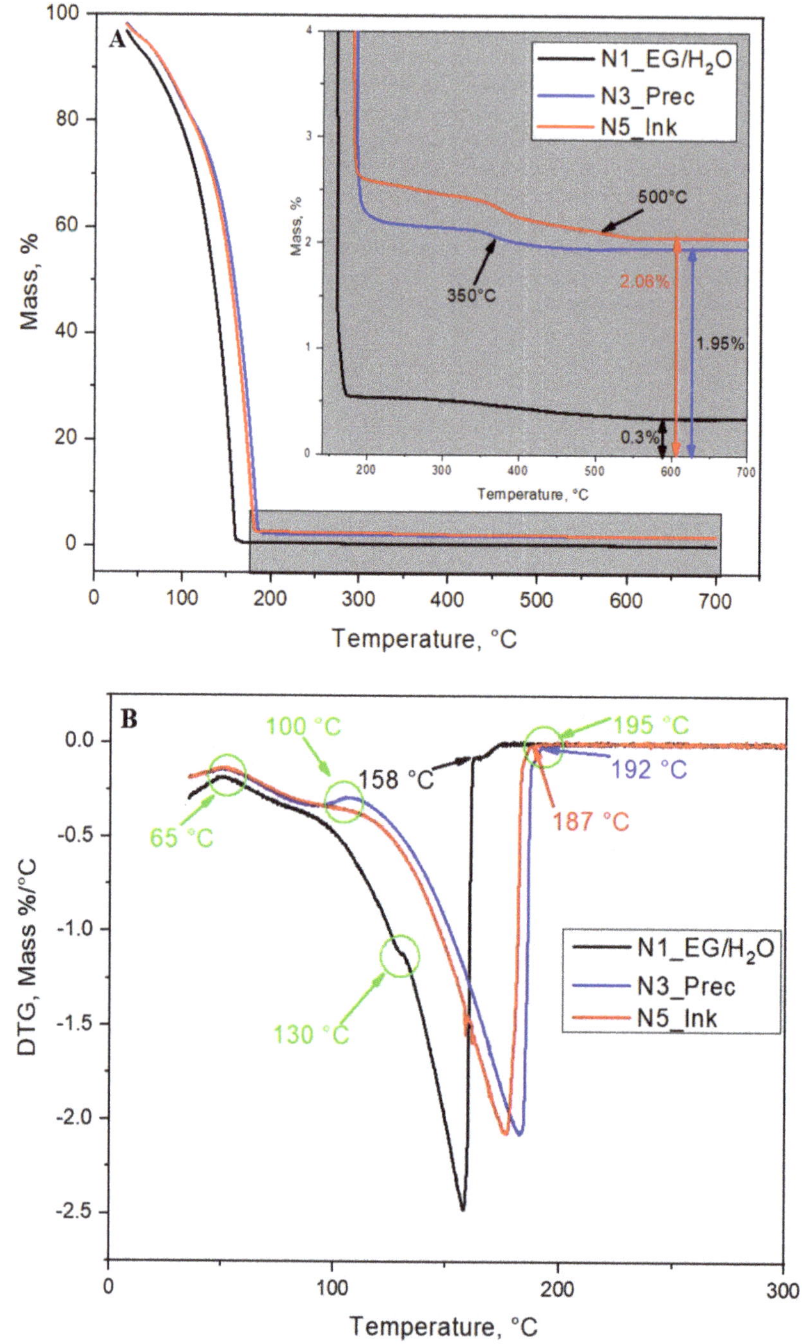

Figure 3. (**A**) The data of TG analysis: the gray insertion demonstrates the peculiarities of the high-temperature tails. (**B**) The data of DTG analysis: the green arrows point to the hidden features in TG data.

The thermogravimetric (TG) and differential thermogravimetric (DTG) analyses show thermal mass loss and its rate during thermal decomposition. For all samples, the interval of 30–200 °C corresponds to the continuous mass loss due to the evaporation of water, EG, and other volatile components. The mass lost for the N1 sample was 99.7% which means that all the EG/H_2O solution was evaporated during the heat treatment. For dispersions N3 and N5, the mass loss during the TG analysis was 98.05% and 97.94%, respectively. Consequently, about 2% of solid sediments remained after the heat treatment of each of the dispersions (see gray insertion in Figure 3A). These values are positively correlated with the theoretical estimate (1.76 wt%) of the NPs mass contained in the AuNCP ink. The displacement of the evaporation point for dispersions N3 and N5 to higher temperatures than N1 and the presence of inflections in the temperature range of 300 and 500 °C is due to the presence of solid nanoscale formations, from which the removal of volatile components requires more energy. The DTG analysis revealed several temperature regions at temperatures about 65, 100, 130, and 195 °C, which are highlighted with green circles in Figure 3B, and are associated with the predominance and overlap of the evaporation processes of various volatile components, the reduction of AuNP, and the formation of AuCNP in the N3 and N5 dispersions. These temperatures were considered for developing the film annealing regime presented in Figure 4.

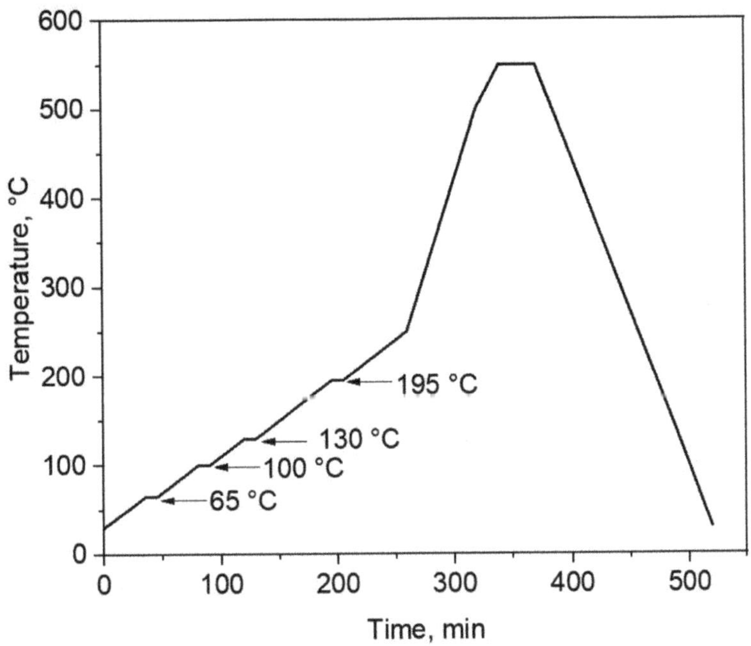

Figure 4. The annealing thermogram for the AuCNP ink.

Based on the ongoing processes, the presented thermogram can be divided into four ranges. In the temperature range from 30 to 100 °C, including 10 min stabilization shelves at 65 and 100 °C, the process of water evaporation prevails. In the second temperature range (from 100 to 195 °C) with stabilization shelves at 130 and 195 °C, the evaporation of EG/H_2O solution and other volatile composites occurs, and the TiO_2-supported Au reduction processes continues. In the third temperature range (from 195 to 550 °C with a 30 min shelf at 550 °C), the final removal of volatile residues and the formation and annealing of the TiO_2 matrix with embedded AuNPs is completed. Note that the entire temperature treatment process of the films was carried out with continuous blowing of the furnace chamber with dry air at a flow rate of 10 sccm. The heating rate was 1

°C/min from 30 to 250 °C, then it was increased fivefold for the range of 250 to 500 °C and finally maintained at 2 °C/min at temperatures from 500 to 550 °C. The cooling rate was 5 °C/min.

In order to elucidate the morphology and size of the particles in the ink, as well as the particles forming on a substrate during drying and annealing, the AuCNP ink was dripped on Cu grids and treated at different temperatures. The STEM images of drops of ink on copper grids after different treatments, i.e., dried at room temperature, heated to 200 °C, and annealed at 550 °C, are presented in Figure 5. We emphasize that the temperature treatment of the samples was carried out in the same furnace, under the same blowdown modes, and according to the corresponding stages shown in the thermogram (Figure 4).

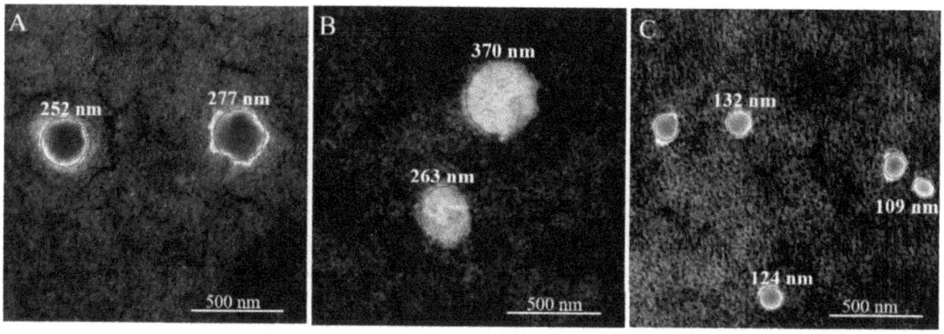

Figure 5. STEM images of the AuCNP evolution in the process of different thermal treatments: (**A**)—drying at room temperature, (**B**)—200 °C heating, (**C**)—annealing at 550 °C.

The diameter of AuCNPs dried at room temperature (Figure 5A), is in the range obtained from the DLS measurements. Their shape was not spherical but more like a drop spreading over a surface. Moreover, it can be seen that the adsorption contrast of the inner parts of AuCNP particles is not uniform, so they actually look like merged, smaller NPs. After heating at 200 °C and evaporation of EG/H_2O (Figure 5B), there was no significant change in the diameter of the AuCNPs. In this case, we registered bright scattered signals over all the surfaces of the particles, which look like an inhomogeneous 2D agglomerate of smaller AuNPs. During the annealing at 550 °C (Figure 5C), the agglomerated NPs merged and formed globular AuCNPs with an average diameter of about 100 nm. In all three images, the AuNPs are embedded in a TiO_2 film, the lower absorption contrast of which, in comparison with the AuNPs, causes its darker display on STEM images. During the subsequent heat treatment, volatile components are removed, and the film becomes more compact and thinner, and, as a result, its signal (Figure 5C) becomes more intense.

3.2. *Characterization of TiO_2/AuNP Films*

Three layers of AuCNP ink sequentially, one on top of the other, were inkjet printed onto a glass substrate. After the printing of each layer, it was dried for two minutes at room temperature. Then, this composite layer was temperature treated according to the thermogram (Figure 4) to form a TiO_2/AuNP film with a thickness of 120 nm (see Figure S3). The thickness of the films was estimated from the cross-sectionional SEM data (see Figure S4).

The XRD analysis of the TiO_2/AuNP films is presented in Figure 6. The synthesized coatings were continuous homogeneous transparent polycrystalline films.

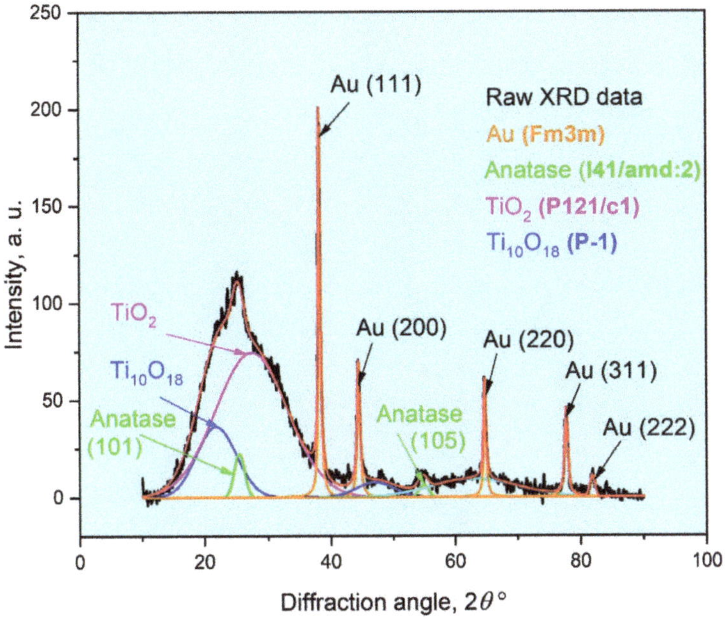

Figure 6. XRD pattern from TiO_2/AuNP film inkjet printed with 7-day-aged ink on a glass substrate and annealed at 550 °C. The black line is the raw X-ray data. The red line is the fitted data as a result of the peak deconvolution analysis. The blue, green, pink, and orange lines present the peaks that belong to $Ti_{10}O_{18}$, Anatase, TiO_2, and Au (Fm3m) phases, respectively.

A pronounced halo in the low-angle region and a peak at the position $2\theta = 25°371'$ correspond to a superposition of diffraction reflections from three phases of titanium oxide: $Ti_{10}O_{18}$, TiO_2, and anatase constituting 21, 75, and 4% of the matrix, respectively. The $Ti_{10}O_{18}$ and TiO_2 phases are ultradisperse with a coherent scattering region (CSR) of 6 Å and 12 Å, respectively, the CSR of the anatase phase is 50 Å. Sharp, well-defined peaks in the diffractogram correspond, in angular positions, to the fcc gold phase. The CSR of this phase is 127 Å. Thus, according to the X-ray diffraction analysis data, the TiO_2/AuNP film consists of three TiO_2 phases, two of which are ultradisperse, and the fcc phase of AuNP. The aging stability of the ink and, as a consequence, the reproducibility of the printed films is the cornerstone of their applicability. The XRD analysis did not reveal a significant difference in the films printed with 1-, 2-, and 7-day-aged inks. The corresponding diffractograms are presented in Figure 7.

A full-profile analysis of the 7 day diffractogram is shown above in Figure 6. Note that the positions, shapes, and intensities of both the TiO_2 halo and the gold peaks did not change with the aging of the ink. More informative in this regard were the results of SEM studies of the morphology of TiO_2/AuNP films. The SEM images of the surface of TiO_2/AuNP films printed with 1-, 2-, 7-, 10-, and 19-day-aged inks and labeled D1, D2, D7, D10, and D19, respectively, are shown in Figure 8. It has been found that the age of the AuCNP ink, meaning how many days have passed since the ink was prepared, is a critical parameter affecting the AuNP size distribution in the TiO_2/AuNP films.

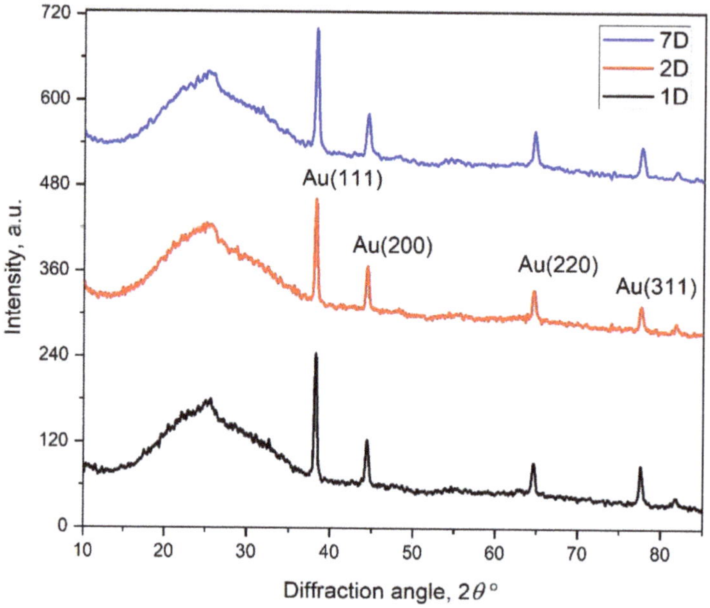

Figure 7. The XRD data of TiO_2/AuNP films printed with 1-, 2-, and 7-day-aged AuCNP inks.

The agglomerates of AuNPs with mean sizes of 80 and 40 nm and a standard deviation (SD) of 30 nm were detected in the D1 and D2 films, respectively. Aging of the ink within seven days of preparation or more led to a significant decrease in the SD and an improvement of the AuNPs size distribution in the TiO_2/AuNP films. The AuNP particles with mean sizes of 63, 29, and 33 nm with a SD of 15 nm were revealed in the films D7, D10, and D19, respectively. We emphasize that, within the measurement error, the mean sizes of AuNPs particles in TiO_2/AuNP films printed with 10 and 19 day ink did not change and, on average, were equal to 30 nm. The size distribution and SD of the AuNP in the TiO_2/AuNP films for the inks at the different storage time, restored from the corresponding SEM images, are presented in Figure 8. Importantly, note that regular monitoring of ink during storage and immediately before printing did not reveal sedimentation processes in them. The HDD of AuCNP particles in dispersion did not change, and the ink remained stable even four months after synthesis (see Supplementary Materials, Figure S5).

The absorption spectra of the TiO_2 and TiO_2/AuNP films printed with TNP dispersion and 10-day-aged inks, respectively, and spherical-shaped AuNPs in water (W/AuNP) [40] are shown in Figure 9. The absorption at 571 nm in TiO_2/AuNP film and 543 nm in W/AuNP are explained by localized surface plasmon resonance (LSPR) of AuNPs.

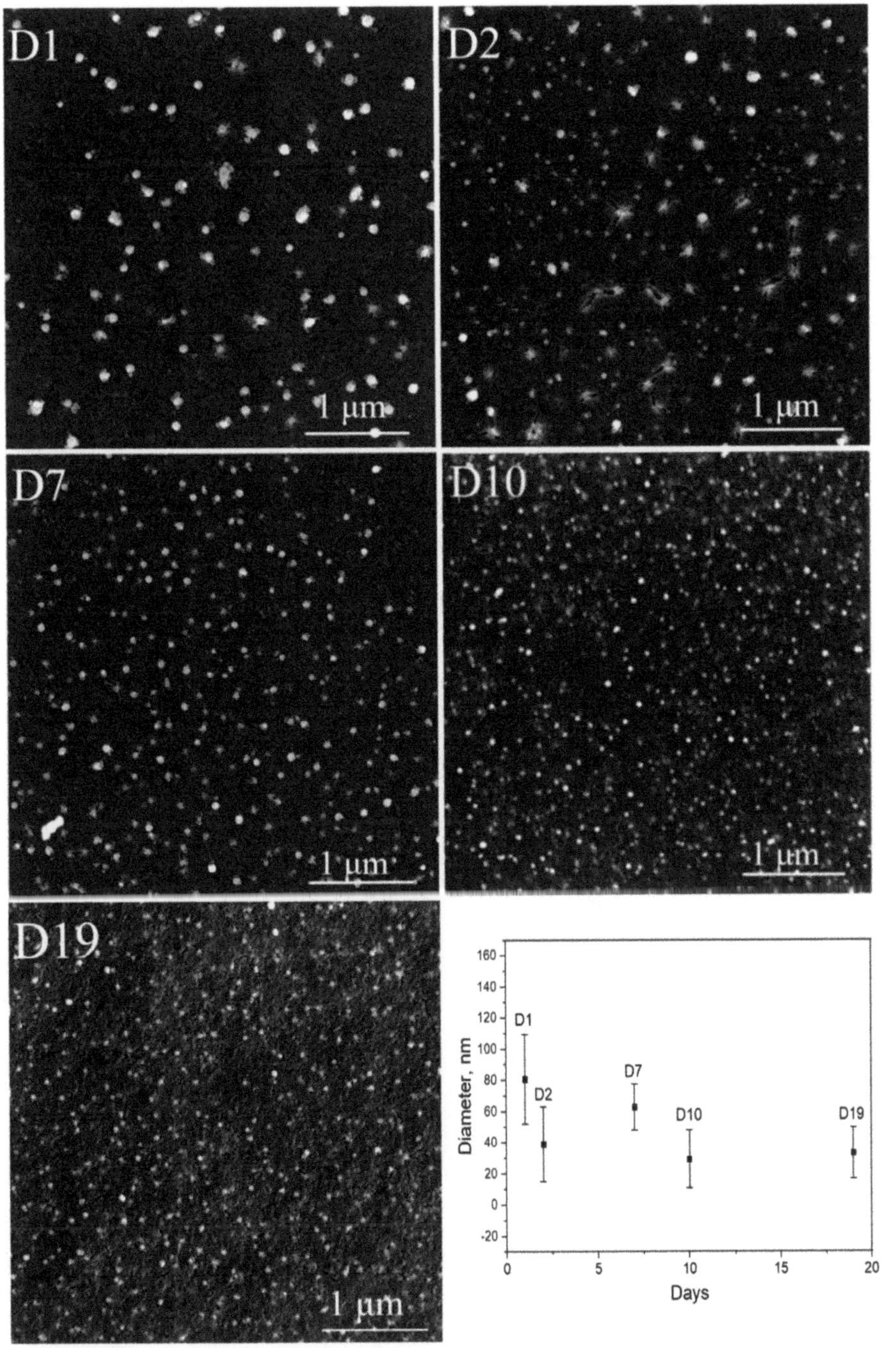

Figure 8. SEM images of the TiO_2/AuNP films printed with the inks of different day-age and the day-age dependence of the size distribution of AuNPs in these films. The bars in the plot correspond to the standard deviation of the size. D1, D2, D7, D10 and D19 on subfigures correspond to 1-, 2-, 7-, 10-, and 19-day-aged inks.

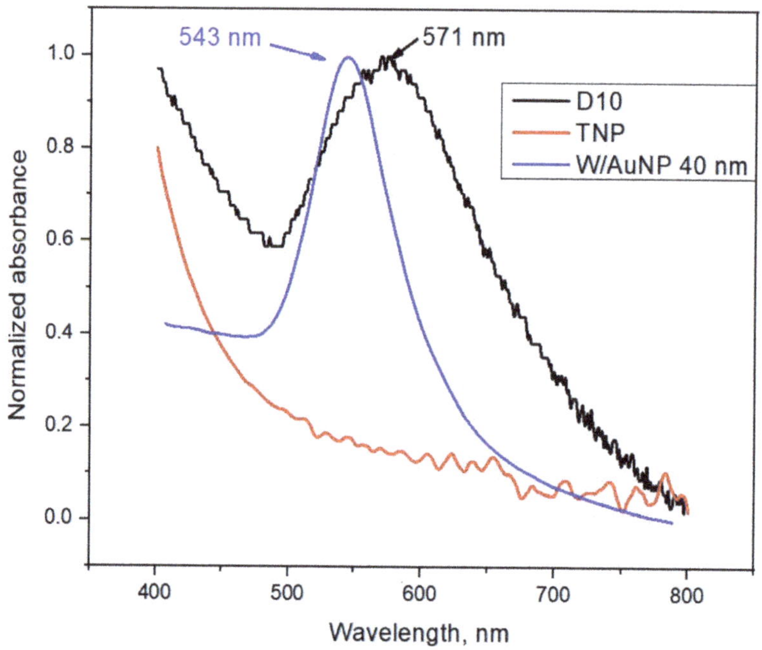

Figure 9. Normalized absorbance spectra of TiO_2, D10 TiO_2/AuNP ink-jet printed films, and 40 nm AuNPs dispersed in water [40].

The peak position of plasmonic absorption is sensitive to the size of AuNPs and the surrounding dielectric environment [41,42]. The average diameter of the AuNPs in the D10 TiO_2/AuNP film is 29 nm (see the plot in Figure 8); however, the corresponding plasmon absorption peak is shifted to the red region, relative to the one of the W/AuNP dispersion, the average particle size of which is 40 nm. This shift is caused by the fact that AuNPs in D10 TiO_2/AuNP films are surrounded by the TiO_2 matrix, whose dielectric constant [43] is higher than that of water. The AuNPs are embedded in the TiO_2 matrix and cause the LSPR absorption of light by the TiO_2/AuNP films in the yellow-green wavelength range. Moreover, by varying the AuNP size, shape, and dielectric environment, this range can be tuned to improve SC absorption.

4. Conclusions

A new strategy for depositing thin TiO_2 films with embedded AuNPs for SC applications was proposed. The films were inkjet printed with AuCNP ink on glass substrates. The method of polyol synthesis of gold nanoparticles on a metal oxide support was modified and optimized to produce the AuCNP ink, which is a dispersion of AuCNP and TNP in the mixture of EG and water. Despite the ink stability, it was revealed, that the ink age factor was critical for the AuNPs size distribution in the TiO_2 matrix. In order to obtain the AuNP imbedded in TiO_2/AuNP film with a mean size of 30 nm and minimal standard deviation, it was necessary to store the ink for at least 10 days after the synthesis. The TiO_2/AuNP films demonstrated a LSPR absorption in the yellow-green wavelength range. The proposed strategy opens up new possibilities for modifying the architecture of perovskite solar cells by introducing LSPR structures. That is a promising approach to increase the efficiency of photovoltaic devices.

Supplementary Materials: The following are available online at https://www.mdpi.com/article/10.3390/coatings11121525/s1, Figure S1: The images of the samples N1, N2, N3, and N5, Figure S2: Particle size distribution of colloidal TNP obtained by DLS, Figure S3: The image of the TiO_2/AuNP film ink-jet printed with 10-day-aged AuCNP ink on a glass substrate, Figure S4: The SEM images of the cross-section of the TiO_2/AuNP film, Figure S5: The day-age dependence of the AuCNP nanoparticles' HDD. The bars on the plot correspond to standard deviation. Reference [44] is cited in the supplementary materials.

Author Contributions: Methodology, S.R., A.M. and V.D.; experiment, S.R.; data analysis, S.R., V.D., A.M. and A.K.; supervision of the project, M.Z. All authors discussed the results and contributed to the final manuscript. All authors have read and agreed to the published version of the manuscript.

Funding: This research received no external funding.

Institutional Review Board Statement: Not applicable.

Informed Consent Statement: Not applicable.

Data Availability Statement: Data sharing is not applicable to this article.

Acknowledgments: The authors are deeply grateful to Yelena Yadgarov for helpful advice and fruitful discussions of the results, Natali Litvak for the SEM images, Alexey Kossenko for the XRD studies, and Michal Weitman (Bar Ilan University) for the TGA measurements. The research work of the student was carried out with the scholarship support of Yuri Dombrovsky.

Conflicts of Interest: The authors declare no conflict of interest.

References

1. Suchomel, P.; Kvitek, L.; Prucek, R.; Panacek, A.; Halder, A.; Vajda, S.; Zboril, R. Simple size-controlled synthesis of Au nanoparticles and their size-dependent catalytic activity. *Sci. Rep.* **2018**, *8*, 4589. [CrossRef] [PubMed]
2. Hu, X.; Zhang, Y.; Ding, T.; Liu, J.; Zhao, H. Multifunctional Gold Nanoparticles: A Novel Nanomaterial for Various Medical Applications and Biological Activities. *Front. Bioeng. Biotechnol.* **2020**, *8*, 990. [CrossRef] [PubMed]
3. Guo, Y.; He, X.; Liu, X.; Li, X.; Kang, L. One-step implementation of plasmon enhancement and solvent annealing effects for air-processed high-efficiency perovskite solar cells. *J. Mater. Chem. A* **2018**, *6*, 24036. [CrossRef]
4. Li, C.; Xu, C.; Cahen, D.; Jin, Y. Unprecedented efficient electron transport across Au nanoparticles with up to 25-nm insulating SiO_2-shells. *Sci. Rep.* **2019**, *9*, 18336. [CrossRef]
5. Yao, K. Plasmonic Metal Nanoparticles with Core-Bishell Structure for High-Performance Organic and Perovskite Solar Cells. *ACS Nano* **2019**, *13*, 5397–5409. [CrossRef] [PubMed]
6. Zhang, C.; Luo, Q.; Shi, J.; Yue, L.; Wang, Z.; Chena, X.; Huang, S. Efficient perovskite solar cells by combination use of Au nanoparticles and insulating metal oxide. *Nanoscale* **2017**, *9*, 2852–2864. [CrossRef]
7. Tran, Q.N.; Lee, H.K.; Kim, J.H.; Park, S.J. Influence of Gold-Silver Rough-Surface Nanoparticles on Plasmonic Light Scattering in Organic Solar Cells. *J. Nanosci. Nanotechnol.* **2020**, *20*, 304–311. [CrossRef]
8. Brown, M.D.; Suteewong, T.; Kumar, R.S.S.; D'Innocenzo, V.; Petrozza, A.; Lee, M.M.; Wiesner, U.; Snaith, H.J. Plasmonic Dye-Sensitized Solar Cells Using Core-Shell Metal-Insulator Nanoparticles. *Nano Lett.* **2011**, *11*, 438–445. [CrossRef]
9. Moakhar, R.S. Recent Advances in Plasmonic Perovskite Solar Cells. *Adv. Sci.* **2020**, *7*. [CrossRef]
10. Wang, B. Enhancing the Photovoltaic Performance of Perovskite Solar Cells Using Plasmonic Au@Pt@Au Core-Shell Nanoparticles. *Nanomaterials* **2019**, *9*, 1263. [CrossRef]
11. Sreedharan, R.S.; Kavitha, V.S.; Suresh, S.; Krishnan, R.R.; Bose, R.J.; Pillai, V.P.M. Tailoring the properties of zinc oxide films by incorporating gold nanoparticles using RF magnetron sputtering. *Appl. Phys. A* **2018**, *124*, 815. [CrossRef]
12. Li, J.; Zhou, H.; Qian, S.; Liu, Z.; Feng, J.; Jin, P.; Liu, X. Plasmonic gold nanoparticles modified titania nanotubes for antibacterial application. *Appl. Phys. Lett.* **2014**, *104*, 261110. [CrossRef]
13. Tan, B.J.Y.; Sow, C.H.; Koh, T.S.; Chin, K.C.; Wee, A.T.S.; Ong, C.K. Fabrication of Size-Tunable Gold Nanoparticles Array with Nanosphere Lithography, Reactive Ion Etching, and Thermal Annealing. *J. Phys. Chem. B* **2005**, *109*, 11100–11109. [CrossRef]
14. Bi, K. Direct electron-beam patterning of transferrable plasmonic gold nanoparticles using a $HAuCl_4$/PVP composite resist. *Nanoscale* **2019**, *11*, 1245. [CrossRef]
15. Keller, K.; Khramenkova, E.V.; Slabov, V.; Musin, A.; Kalashnikov, A.; Vinogradov, A.V.; Pidko, E.A. Inkjet Printing of Sc-Doped TiO_2 with Enhanced Photoactivity. *Coatings* **2019**, *9*, 78. [CrossRef]
16. Turkevich, J.; Stevenson, P.C.; Hiller, J. A study of the nucleation and growth processes in the synthesis of colloidal gold. *Discuss. Faraday Soc.* **1951**, *11*, 55–75. [CrossRef]
17. Kimling, J.; Maier, M.; Okenve, B.; Kotaidis, V.; Ballot, H.; Plech, A. Turkevich Method for Gold Nanoparticle Synthesis Revisited. *J. Phys. Chem. B* **2006**, *110*, 15700–15707. [CrossRef]

18. Brust, M.; Walker, M.; Bethell, D.; Schiffrin, D.J.; Whyman, R. Synthesis of Thiol-derivatised Gold Nanoparticles in a Two-phas Liquid-Liquid System. *J. Chem. Soc. Chem. Commun.* **1994**, *1994*, 801–802. [CrossRef]
19. Deraedt, C.; Salmon, L.; Gatard, S.; Ciganda, R.; Hernandez, R.; Ruiza, J.; Astruc, D. Sodium borohydride stabilizes very activ gold nanoparticle catalysts. *Chem. Commun.* **2014**, *50*, 14194–14196. [CrossRef]
20. Fiévet, F. The polyol process: A unique method for easy access to metal nanoparticles with tailored sizes, shapes and composition *Chem. Soc. Rev.* **2018**, *47*, 5187–5233. [CrossRef]
21. Silvert, P.Y.; Tekaia-Elhsissen, K. Synthesis of monodisperse submicronic gold particles by the polyol process. *Solid State Ionic* **1995**, *82*, 53–60. [CrossRef]
22. Sankar, M. Role of the Support in Gold-Containing Nanoparticles as Heterogeneous Catalysts. *Chem. Rev.* **2020**, *120*, 3890–393 [CrossRef] [PubMed]
23. Buonerba, A.; Grassi, A. Trends in Sustainable Synthesis of Organics by Gold Nanoparticles Embedded in Polymer Matrice *Catalysts* **2021**, *11*, 714. [CrossRef]
24. Haruta, M.; Tsubota, S.; Kobayashi, T.; Kageyama, H.; Genet, M.J.; Delmon, B. Low-Temperature Oxidation of CO Over Gol Supported on TiO_2, α-Fe_2O_3, and Co_3O_4. *J. Catal.* **1993**, *144*, 175–192. [CrossRef]
25. Soejima, T.; Tada, H.; Kawahara, T.; Ito, S. Formation of Au Nanoclusters on TiO_2 Surfaces by a Two-Step Method Consisting c Au(III)-Complex Chemisorption and Its Photoreduction. *Langmuir* **2002**, *18*, 4191–4194. [CrossRef]
26. Aguirre, M.E.; Perelstein, G.; Feldhoff, A.; Condó, A.; Tolley, A.J.; Grela, M.A. The spontaneous room temperature reduction c $HAuCl_4$ in ethylene glycol in the presence of ZnO: A simple strategy to obtain stable Au/ZnO nanostructures exhibiting stron surface plasmon resonance and efficient electron storage properties. *New J. Chem.* **2015**, *39*, 909. [CrossRef]
27. Pereira, J.M.d.S.; Ciotti, L.; Vaz, J.M.; Spinacé, E.V. Preparation of Au/TiO_2 Catalyst by a Liquid-Phase Reduction Method fo Preferential Oxidation of Carbon Monoxide in a Hydrogen Rich-Stream (CO-PROX reaction). *Mater. Res.* **2018**, *21*. [CrossRef]
28. Haruta, M. Size- and support-dependency in the catalysts of gold. *Catal. Today* **1997**, *36*, 153–166. [CrossRef]
29. Haruta, M. Role of perimeter interfaces in catalysis by gold nanoparticles. *Faraday Discuss.* **2011**, *152*, 11–32. [CrossRef]
30. Brar, K.S.; Verma, M. Measurement of nanoparticles by light-scattering techniques. *Trends Anal. Chem.* **2011**, *30*, 4–17. [CrossRef
31. Xu, R. Progress in nanoparticles characterization: Sizing and zeta potential measurement. *Particuology* **2008**, *6*, 112–115. [CrossRef
32. Murphy, P.J.; LaGrange, M.S. Raman spectroscopy of gold chloro-hydroxy speciation in fluids at ambient temperature an pressure: A re-evaluation of the effects of pH and chloride concentration. *Geochim. Cosmochim. Acta* **1998**, *62*, 3515–352 [CrossRef]
33. Ivanova, S.; Petit, C.; Pitchon, V. A new preparation method for the formation of gold nanoparticles on an oxide support. *App Catal. A Gen.* **2004**, *267*, 191–201. [CrossRef]
34. Zanella, R.; Giorgio, S.; Henry, C.R.; Louis, C. Alternative Methods for the Preparation of Gold Nanoparticles Supported on TiO_2 *J. Phys. Chem. B* **2002**, *106*, 7634–7642. [CrossRef]
35. Mazza, T.; Barborini, E.; Piseri, P.; Milani, P.; Cattaneo, D.; Li Bassi, A.; Ducati, C. Raman spectroscopy characterization ofTiO_2 rutile nanocrystals. *Phys. Rev. B* **2007**, *75*, 045416. [CrossRef]
36. Krishnan, K.; Krishnan, R.S. Raman and infrared spectra of ethylene glycol. *Proc. Indian Acad. Sci. Sect. A* **1966**, *64*, 111–12 [CrossRef]
37. Hoath, S.D. Fundamentals of Inkjet Printing. In *The Science of Inkjet and Droplets*; Wiley: Weinheim, Germany, 2016; p. 9 [CrossRef]
38. Reis, N.; Derby, B. Ink jet deposition of ceramic suspensions: Modelling and experiments of droplet formation. *Mater. Res. Soc Symp. Proc.* **2000**, *625*, 117–122. [CrossRef]
39. Jang, D.; Kim, D.; Moon, J. Influence of fluid properties on ink-jet printability. *Langmuir* **2009**, *25*, 2629–2635. [CrossRef] [PubMed
40. Kim, D.-K.; Hwang, Y.J.; Yoon, C.; Yoon, H.-O.; Chang, K.S.; Lee, G.; Lee, S.; Yi, G.-R. Experimental approach to the fundamenta limit of the extinction coefficients of ultra-smooth and highly spherical gold nanoparticles. *Phys. Chem. Chem. Phys.* **2015**, *17* 20786–20794. [CrossRef]
41. Lavie, A.; Yadgarov, L.; Houben, L.; Popovitz-Biro, R.; Shaul, T.E.; Nagler, A.; Suchowski, H.; Tenne, R. Synthesis of core–shel single-layer MoS_2 sheathing gold nanoparticles, AuNP@ 1L-MoS_2. *Nanotechnology* **2017**, *28*, 24LT03. [CrossRef]
42. Aiboushev, A.; Gostev, F.; Shelaev, I.; Kostrov, A.; Kanaev, A.; Museur, L.; Traore, M.; Sarkisova, O.; Nadtochenko, V. Spectra properties of the surface plasmon resonance and electron injection from gold nanoparticles to TiO_2 mesoporous film: Femtosecond study. *Photochem. Photobiol. Sci.* **2012**, *12*, 631–637. [CrossRef] [PubMed]
43. Bonkerud, J.; Zimmermann, C.; Weiser, P.M.; Vines, L.; Monakhov, E.V. On the permittivity of titanium dioxide. *Sci. Rep.* **2021**, *11* 12443. [CrossRef] [PubMed]
44. Behnajady, M.A.; Eskandarloo, H.; Modirshahla, N.; Shokri, M. Sol-gel low-temperature synthesis of stable anatase-type TiO_2 nanoparticles under different conditions and its photocatalytic activity. *Photochem. Photobiol.* **2011**, *87*, 1002–1008. [CrossRef [PubMed]

coatings

MDPI

Article

The Effect of Adding Sodium Carbonate on the Electrical Conductivity of Aluminum Paste

Shunke Liu, Xiaoyun Zhu * and Jinming Long

School of Materials Science and Engineering, Kunming University of Science and Technology, Kunming 650093, China; skliu1010@163.com (S.L.); jmlong7880@163.com (J.L.)
* Correspondence: xyzhu@kust.edu.cn

Abstract: Base metal pastes have been widely used in the preparation of ZnO varistor electrodes, and it is important to accurately grasp the relevant mechanisms affecting the conductivity of aluminum electrodes. In this paper, the effect of adding sodium carbonate on the conductive property of aluminum paste was assessed, and the microscopic mechanism during aluminum electrode sintering explored. The results show that adding sodium carbonate can reduce the softening point of glass powder and enhance its fluidity. Sodium carbonate, glass, and aluminum oxide film react together; consequently, the aluminum oxide film is partially dissolved by reaction to produce defects, and there is tight contact at the interface between the aluminum powder particles. The sodium ions will displace the aluminum ions in the alumina, conferring the alumina film with a certain ionic conductivity. At the same time, sodium ions are doped into the aluminum lattice, which causes the aluminum lattice to swell. After sintering, the structure of aluminum electrode is compact and its electrical conductivity is significantly improved. This study is a valuable reference for the theoretical research and the potential applications of aluminum paste.

Keywords: aluminum paste; conductivity; sodium carbonate

1. Introduction

Conductive paste is an important and indispensable basic raw materials in microelectronics industry. It is mainly used as electrode material for manufacturing electronic components. A zinc oxide varistor is a polycrystalline material [1,2]. It is widely used in circuits for overvoltage protection and voltage regulation and is characterized by a sensitive change in resistance value as the applied voltage increases, resulting in a nonlinear change in the corresponding current and voltage [3,4]. At present, silver and other precious metals are usually used as electrode materials for zinc oxide varistors in the industry. With the rapid development of the electronics and information industry, the demand for low-cost electronic components in the electronics manufacturing industry is more and more extensive. Zinc oxide varistors with silver as the electrode material have been at a disadvantage in the competitive market due to their high costs [5,6]. Hence, cost reduction by the widespread use of base metals for the electrode material has become one of the trends in the development of zinc oxide varistors. Copper electrodes have gradually replaced the silver electrode for preparing zinc oxide varistors [7,8], but the former is prone to oxidization, and the sintering process is also complex (must be carried out in a protective atmosphere), with high costs [9–11]. Therefore, the development of a base metal electrode paste with good performance and low cost (that can be sintered in air) has become a topic requiring research.

The cost of aluminum electrode material is lower, but the chemical properties of aluminum powder of the conductive phase of aluminum paste confer greater activity. Usually, there is a layer of insulating alumina film on the surface, which leads to the poor electrical conductivity of the aluminum electrode [12–14]. Even if the aluminum powder is pretreated (removing the alumina film), it will still oxidize in the high temperature environment during sintering, resulting in reduced performance. In this paper, aluminum paste was prepared with the addition of different contents of sodium carbonate to study its effect on the performance of aluminum electrode, and the underlying mechanism, so as to provide a scientific basis for improving the performance of aluminum electrodes.

2. Materials and Methods

Experimental materials: spherical aluminum powder, particle size of 2.0–3.0 μm; the glass was SiO_2-B_2O_3-Bi_2O_3 system glass. The organic carrier was a mixture of butyl carbitol, ethyl cellulose, and other additives.

Aluminum powder, glass powder, organic carrier, and sodium carbonate were weighed in a certain proportion, and then rolled and dispersed with a three-roller mill to obtain aluminum paste with fineness of less than 10 μm. The aluminum paste was printed on ZnO substrate by a screen printing method, placed into an oven and dried at 150 °C for 5 min, and finally transferred into a tube sintering furnace (NBD-HT1100-80IT, Henan Nobody Materials Science and Technology Co., Ltd., Zhengzhou, China) for sintering. The peak temperature was 620 °C, the holding time was 10 min, and the heating rate was 30 °C/min. The content of sodium carbonate (mass fraction, %) added to the aluminum pastes were 0%, 3%, 4%, 5%, and 6% for 5 sets of samples. The sintered aluminum electrode specimens are shown in Figure 1, and the sintering curves are shown in Figure 2.

Figure 1. Sintered aluminum electrode sample.

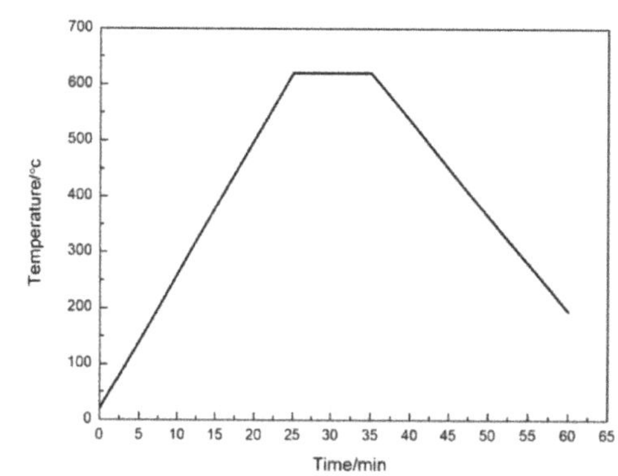

Figure 2. Sintering curve of aluminum electrode at 620 °C in air.

The microstructure and densities of aluminum electrodes were characterized by scanning electron microscopy (SEM, FEI Company, Hillsboro, OR, USA) XL30ESEM-

TMP. The square resistance of the aluminum electrode was measured by the four-probe method. To distinguish the metallic core of the aluminum powder from the oxide shell, the evolution of the powder morphology and core-shell structure after sintering was studied using a Tecnai F30 transmission electron microscope (TEM, Hillsboro, OR, USA). The physical phase analysis of the aluminum electrode as well as the glass powder was checked using a D8A advance x-ray diffractometer. Possible phase changes in glass, sodium carbonate, and mixtures of glass and sodium carbonate were characterized by thermogravimetry (TG) and differential scanning calorimetry (DSC) STA 449F3. The samples were heated in an oxygen/argon atmosphere (oxygen flow rate of 20 mL/min and argon flow rate of 80 mL/min) from room temperature to 700 °C at a heating rate of 10 °C/min. XPS analysis was performed using a PHI5000 Versaprobe-II X-ray photoelectron spectrometer (ULVAC-PHI, Chigasaki, Japan). The C, Al, and Na elements of the surface of aluminum electrode were measured. C1s (284.80eV) of polluted carbon was used for charge calibration. A survey scan was performed within the range of 0~1400 eV, followed by a narrow scan of C, Al, and Na elements.

3. Materials and Methods

3.1. TG-DSC Analysis of Samples

Figure 3 shows the TG-DSC curves of glass, sodium carbonate, glass and sodium carbonate mixture; Figure 4 shows the XRD patterns of glass before and after sintering, and glass and sodium carbonate mixture after sintering. The structural state of glass will change with the increase of temperature during heating. Glass transition is an endothermic process, while crystallization is an exothermic process, so the temperature in the curve rises from low to high, and different endothermic and exothermic peaks correspond to the transition temperature (T_g), softening temperature (T_f), and crystallization temperature (T_s) of glass [15]. As can be seen from Figure 3, before sodium carbonate is added, the glass transition temperature, T_g, is 484 °C and the glass softening temperature, T_f, is 574 °C. The starting point of the exothermic peak shows that the crystallization temperature, T_S, is 584 °C, and at this temperature, the crystallization and nucleation of the glass occur simultaneously. For the glass with added sodium carbonate, T_g was 469 °C, T_f was 499 °C, and T_s was 534 °C. Compared with the glass without sodium carbonate, these three characteristic temperatures are all reduced, indicating that the addition of sodium carbonate can enhance glass sintering [16–19]. It can also be seen from Figures 3 and 4 that the sample of glass with sodium carbonate added has lost weight after sintering, and a new phase of $Bi_{24}Si_2O_{40}$ is generated, which is called sillente, a member of class of stable Bi_2O_3 derivatives in which Si can stabilize γ-Bi_2O_3 to a $Bi_{24}Si_2O_{40}$ phase at lower temperatures [20]. In contrast, the diffraction peak of sodium carbonate is not apparent. This indicates that in the temperature range of 429 to 514 °C, sodium carbonate and glass components undergo a double decomposition reaction, resulting in the release of CO_2 gas, and hence a weight loss. Elemental sodium carbonate should begin to decompose only above 850 °C [21,22], but when the sodium carbonate is added into glass, it will react with glass (generating a new $Bi_{24}Si_2O_{40}$ phase), enhance the softening of glass, and causing the remaining sodium carbonate to be decomposed at a lower temperature (619 °C).

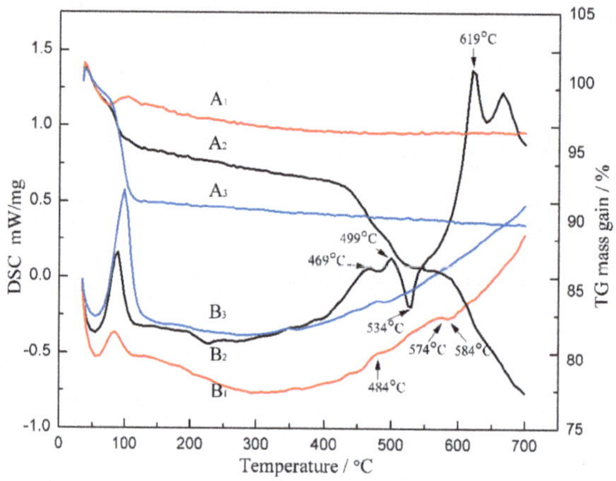

Figure 3. TG−DSC curves of glass, sodium carbonate, and the glass and sodium carbonate mixture. A1, B1: TG−DSC curve of glass, A2, B2: TG−DSC curve of glass and sodium carbonate mixture, A3, B3: TG−DSC curve of sodium carbonate.

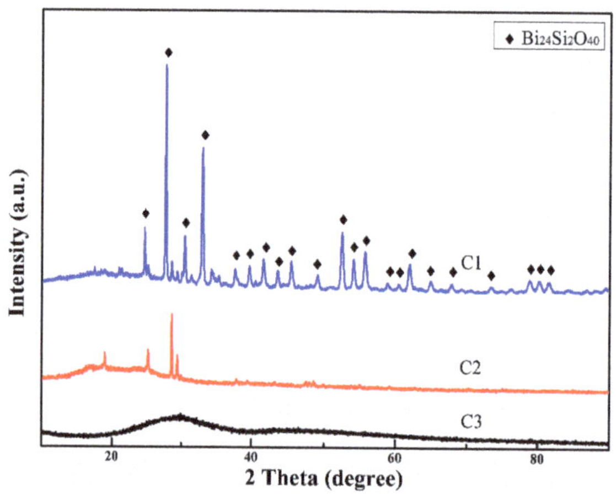

Figure 4. XRD patterns of glass before and after sintering and glass and sodium carbonate mixture after sintering. C1: glass and sodium carbonate mixed powder sintering, C2: Glass after sintering, C3: Unsintered glass.

3.2. XPS Analysis of Samples

Figure 5 shows the XPS spectra of different elements on the surface of the aluminum electrode. Among them, Figure 5a shows the 2p peak of the Al element and Figure 5b shows the 1s peak of the Na element. Without the addition of sodium carbonate, the 2p peak of Al is located at 74.15eV. Since a small amount of sodium carbonate is added as a burning aid during the preparation of glass, a small amount of Na_2CO_3 is also detected in the aluminum electrode sample without added sodium carbonate. Due to the addition of sodium carbonate, the charge transfer changes the environment where Al^{3+} is located, but does not cause a change in valence state. Therefore, the binding energy of Al 2p electrons was reduced to 73.59 eV. From Table 1, it can be seen that the aluminum content in the aluminum electrode without added sodium carbonate was 67.93% and the aluminum content in the aluminum electrode with added sodium carbonate was 29.55%. The large decrease in aluminum content was due to the displacement of aluminum ions in the alumina film by sodium ions. Sodium ion incorporated alumina displays ionic conductivity [23]. The binding energies of Na 1s were 1072.20 eV, 1071.50 eV, and 1070.80 eV, where 1072.20 eV and 1071.50 eV were consistent with the binding energies of Na 1s in the literatur, which proved that Na existed in the form of albite and Na_2CO_3. This also agreed with the XRD test results.

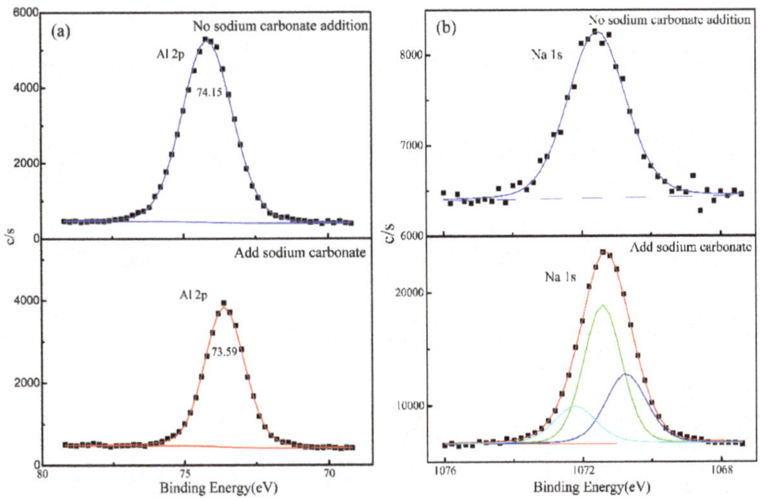

Figure 5. XPS spectra of different elements. (**a**) Al element, (**b**) Na element.

Table 1. Elemental analysis results of aluminum electrodes.

Sample	Content/%		
	C	Al	Na
Aluminum electrode without added sodium carbonate	25.18	67.93	6.89
Aluminum electrode with added sodium carbonate	27.38	29.55	43.07

3.3. TEM and XRD Analysis of Samples

Figure 6 shows the XRD patterns of aluminum paste sintered samples with and without the addition of sodium carbonate. Against the standard PDF cards (JCPDS Card No.49-0003, No.49-0006, No.83-1466) and compared with the samples without sodium carbonate, the diffraction curves of $Na_{1.95}Al_{1.95}Si_{0.05}O_4$, $Na_{1.55}Al_{1.55}Si_{0.45}O_4$, and $Na_{0.986}(Al_{1.005}Si_{2.995}O_8)$ composite oxides appear in the sample with added sodium carbonate. This shows that during the sintering of aluminum paste, in addition to the chemical reaction between the sodium carbonate mentioned above and the glass, there is the formation of a new phase ($Bi_{24}Si_2O_{40}$). The morphology of the aluminum powder after sintering at 620 °C is shown in Figure 7a. EDS analysis was performed on point 1 in Figure 7a, and the results are shown in Figure 7b. The copper mesh was used as the carrier mesh for the TEM test, so the copper diffraction peaks appeared in the EDS analysis. Sodium carbonate, glass, and aluminum oxide film react together, so that the aluminum oxide film is partially dissolved by reaction to produce defects (see Figure 8). Further, the sodium ions will displace the aluminum ions in the alumina, so that the alumina film gains a certain ionic conductivity. As a result, the contact between the aluminum powder particles becomes denser, the contact resistance reduced, and the square resistance of aluminum electrodes is improved.

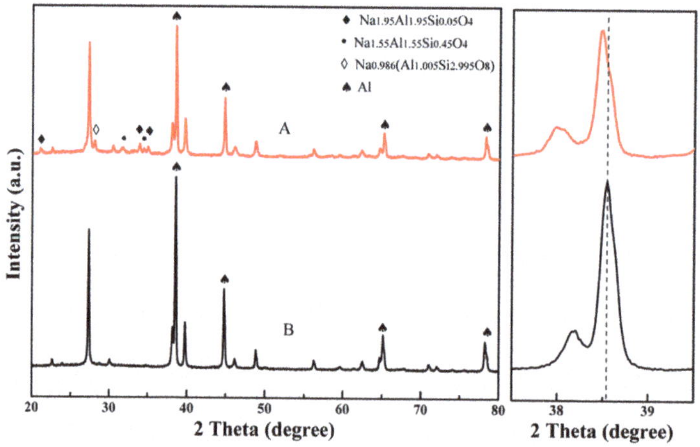

Figure 6. XRD patterns of samples with and without sodium carbonate. A: Sample with sodium carbonate added; B: Sample without sodium carbonate added.

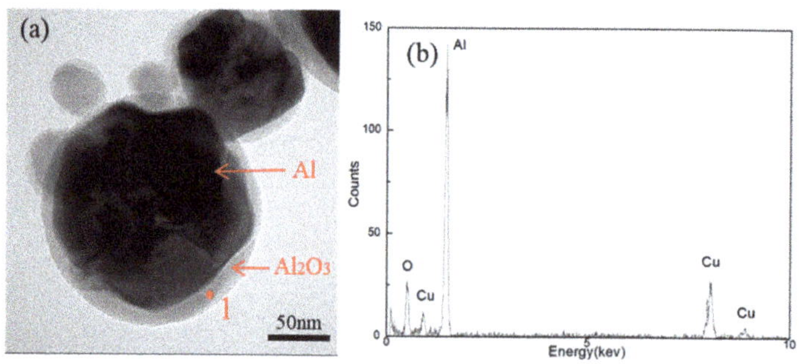

Figure 7. TEM image and EDS analysis of aluminum powder after sintering at 620 °C. (**a**): TEM image of aluminum powder, (**b**): EDS spectrum of point 1 of (**a**).

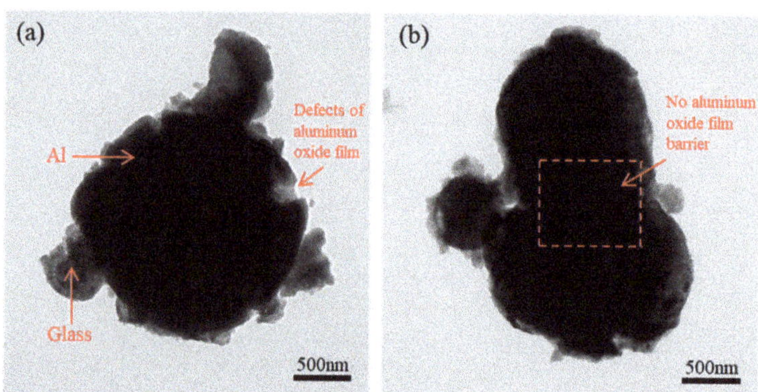

Figure 8. TEM images of the mixture of aluminum powder, glass powder and sodium carbonate after sintering at 620 °C. (**a**): TEM image of a single aluminum particle in the mixed powder; (**b**): TEM image of the connection between aluminum particles in the mixed powder.

In addition, the XRD analysis also shows that, compared with the sintered sample without sodium carbonate, the peak of the aluminum diffraction line of the sample with added sodium carbonate shifts to the left (2θ value decreases); that is, the corresponding plane spacing or the aluminum lattice constant increases. The reason for this is that during the sintering process, when the sodium carbonate reacts with glass and the aluminum powder, the sodium ions become doped into the lattice of aluminum. Due to the large radius of sodium ions (102 pm) and the small radius of aluminum ions (53.5 pm), the sodium ions replace the aluminum ions, resulting in lattice expansion.

3.4. SEM Analysis of Aluminum Electrode Surface

Figure 9 shows the SEM photos of the surfaces of the aluminum electrode with 0%, 3%, 4%, 5%, and 6% sodium carbonate added respectively. Figure 10 shows the square resistance of the aluminum electrode with 0%, 3%, 4%, 5%, and 6% sodium carbonate content respectively. When sodium carbonate is added, aluminum paste enhances the melting of glass in the sintering process. At the same time, aluminum oxide and glass react, and part of aluminum oxide detaches after the reaction, exposing aluminum, so that aluminum particles make contact with each other to form a dense structure. As the sodium carbonate content increases, the ability of glass to infiltrate aluminum particles also increases, and the compactness of aluminum electrode is further improved. As can be seen from Figure 9, the surface morphology of the aluminum electrode varies also significantly. The surface of the aluminum electrode without added sodium carbonate has many holes, with no close contacts between particles, which indicates that the glass does not form a network structure around the aluminum particles to make a good compaction. The square resistance of the aluminum electrode (96.51 mΩ/□) is large, and there are many holes. The number of holes on the surface of the aluminum electrode with 3% sodium carbonate added is still large, but there is a shorter bond between particles, the contact area increases, and the square resistance (78.18 mΩ/□) decreases. The number of holes on the surface of the aluminum electrode with 4% sodium carbonate added was significantly reduced, and the aluminum powder with different particle sizes was distributed proportionally. The large and small aluminum powder particles are arranged relatively close, forming a coherent conductive network. The contact resistance is small, the density is increased, and the square resistance (49.56 mΩ/□) is the smallest. When adding excessive sodium carbonate (>4%), sodium carbonate, which is not involved in the reaction, will become resident on the aluminum electrodes. This affects wettability of glass to aluminum particles, increases the number of holes on the aluminum electrodes, and, at the same time may, block direct contact between aluminum powder particles. Moreover, as there is excess decomposed sodium oxide, the mechanical strength, thermal stability, and chemical stability of the glass may be reduced. As can be seen from Figure 10, compared with the aluminum electrode with 4% sodium carbonate, the square resistance of the aluminum electrode with 5% and 6% sodium carbonate added increases in turn (49.56 mΩ/□→56.7 mΩ/□→ 65.33 mΩ/□).

3.5. SEM Analysis of Aluminum Electrode Cross-Section

The cross-sectional SEM photographs of the aluminum electrode film and the ZnO substrate are given in Figure 11. From Figure 11a, it can be seen that the aluminum electrode film layer with 0% sodium carbonate content added to the ZnO substrate is obviously delaminated, the aluminum electrode layer is not dense, and no diffusion permeation phenomenon occurs between the aluminum electrode film layer and the ZnO substrate. From Figure 11b,c, it can be seen that when 3% and 4% sodium carbonate was added to the aluminum electrode, sodium carbonate promoted the melting of the glass and improved the fluidity of the glass. There is no delamination between the film layer of the aluminum electrode and the ZnO substrate, and the connection effect is better. When 5% and 6% sodium carbonate was added to the aluminum electrode, too much sodium carbonate decomposes to sodium oxide, reducing the mechanical strength, thermal stability, and chemical stability of the glass, resulting in the poor fluidity of the glass. In addition, during the sintering process, there is an increase in $NaAlSiO_4$ crystals precipitated by the reaction of sodium carbonate, aluminum powder, and glass, which hinders the connection between aluminum particles and affects the electrical conductivity. From Figure 11d,e, it can be seen that only a small part of the aluminum electrode film layer and the ZnO substrate did not show delamination, and most of the regions showed delamination and poor adhesion between the aluminum electrode film layer and the ZnO substrate.

Figure 9. Surface SEM of aluminum film samples with different added sodium carbonate content. (**a**): 0%, (**b**): 3%, (**c**): 4%, (**d**): 5%, (**e**): 6%.

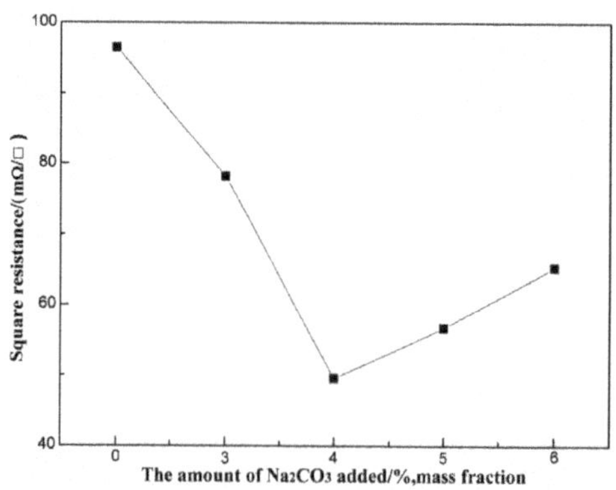

Figure 10. Effect of sodium carbonate content on the square resistance of aluminum electrode specimens.

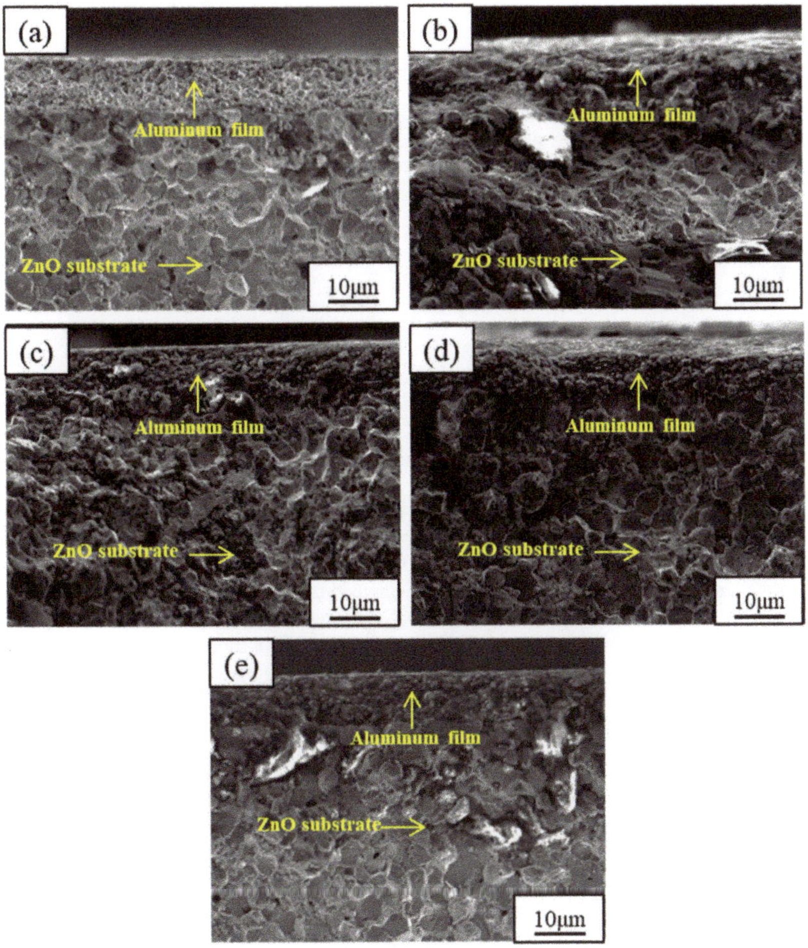

Figure 11. The cross-sectional SEM photographs of aluminum electrode sections with different contents of sodium carbonate added. (**a**): 0%, (**b**): 3%, (**c**): 4%, (**d**): 5%, (**e**): 6%.

3.6. Square Resistance of Aluminum Electrode with Different Sintering Temperatures

It can be seen from Figure 12 that the square resistance of the aluminum electrode decreased substantially with an increase in temperature. At 620 °C, the aluminum electrode conductivity was the best, with a square resistance of 49.56 mΩ/□. With a temperature increase to 640 °C, the square resistance of the aluminum electrode starts to increase due to the enhanced reaction of the glass, aluminum powder, and sodium carbonate, resulting in the production of excess crystals. After the comparison, the best conductivity of the aluminum electrode was obtained from preparation at 620 °C.

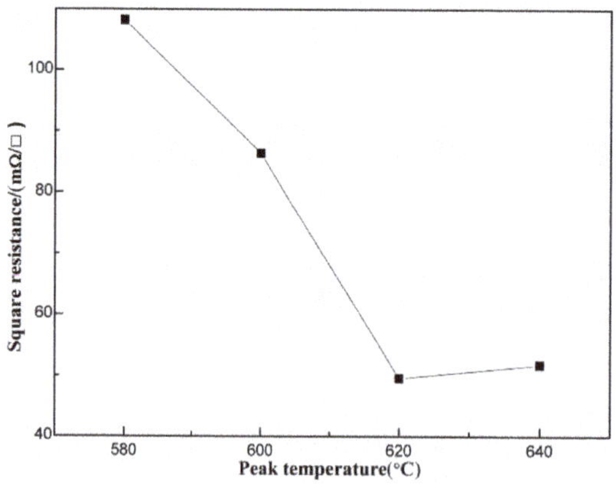

Figure 12. Square resistance curves of aluminum electrodes sintered with 4% sodium carbonate content at different peak temperatures from 580 °C to 640 °C.

3.7. The Effect of the Sodium Carbonate Addition on the Conductivity Mechanism

Figure 13 shows a schematic representation of the effect of the sodium carbonate addition on electrical conductivity, with a clear understanding that this addition process improves the performance of aluminum electrode. As can be seen from the figure, before adding sodium carbonate (Figure 13a), the contact between aluminum particles is blocked by aluminum oxide film, and the electrical conductivity is poor. After adding sodium carbonate (Figure 13b), the melting of the glass powder is enhanced in the sintering process, and the fluidity is better, which is conducive to closer contacts in the aluminum powder. Sodium carbonate, glass, and aluminum oxide film react together, and the aluminum oxide film is partially dissolved by the reaction to produce defects. Sodium ions are doped into the aluminum lattice, and, at the same time, the sodium ions displace the aluminum ions in the alumina, so that the alumina film has a certain ionic conductivity; thus, this reduces the contact resistance between aluminum powder particles and improves electrical conductivity.

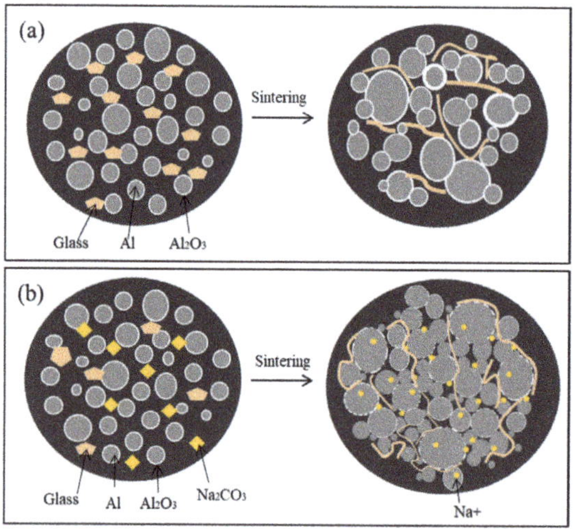

Figure 13. Model diagram of conductive mechanism enhanced by addition of sodium carbonate. (**a**): Sintering without addition of sodium carbonate; (**b**): Sintering with addition of sodium carbonate.

4. Conclusions

During the sintering process of aluminum paste, sodium carbonate reacts with the aluminum oxide film on the surface of glass and aluminum powder, which causes the film to be partially dissolved and removed. The contact between the aluminum powder particles is increased and the square resistance decreases. During the sintering process, sodium ions can diffuse into the surface lattice of the aluminum powder through the location where the alumina film is removed by dissolution, causing the crystal lattice to expand. At the same time, the sodium ions displace the aluminum ions in the alumina, conferring the alumina film with a certain ionic conductivity, which contributes to an improvement in electrical conductivity. The addition of an appropriate amount of sodium carbonate to the aluminum paste reduced the square resistance from 96.51 m$\Omega/\square$ to 49.56 m$\Omega/\square$, by about 50%.

Author Contributions: Data curation, S.L.; Formal analysis, S.L.; Supervision, X.Z. and J.L.; investigation, S.L., X.Z. and J.L.; Writing—original draft, S.L. All authors have read and agreed to the published version of the manuscript.

Funding: This research received no external funding.

Institutional Review Board Statement: Not applicable.

Informed Consent Statement: Not applicable.

Data Availability Statement: Data is contained within the article.

Conflicts of Interest: The authors declare no conflict of interest.

References

1. Xu, D.; Shi, L.; Wu, Z.; Zhong, Q.; Wu, X. Microstructure and electrical properties of ZnO–Bi_2O_3-based varistor ceramics by different sintering processes. *J. Eur. Ceram. Soc.* **2009**, *29*, 1789–1794. [CrossRef]
2. Wang, Q.; Qin, Y.; Xu, G.J.; Chen, L.; Li, Y.; Duan, L.; Li, Z.X.; Li, Y.L.; Cui, P. Low-voltage ZnO varistor fabricated by the solution-coating method. *Ceram. Int.* **2008**, *34*, 1697–1701. [CrossRef]
3. Yuan, K.; Li, G.; Zheng, L.; Cheng, L.; Meng, L.; Yao, Z.; Yin, Q. Improvement in electrical stability of ZnO varistors by infiltration of molten Bi_2O_3. *J. Alloys Comp.* **2010**, *503*, 507–513. [CrossRef]
4. Xu, D.; Shi, X.F.; Cheng, X.N.; Yang, J.; Fan, Y.; Yuan, H.M.; Shi, L.Y. Microstructure and electrical properties of Lu_2O_3-doped ZnO-Bi_2O_3-based varistor ceramics. *Trans. Nonferrous Met. Soc. China* **2010**, *12*, 2303–2308. [CrossRef]
5. Gan, W.P.; Yue, Y.X.; L, L.; Pan, Q.Y.; Xiong, Z.J. Preparation of Lead Free Conductive Silver Paste and Its Sintering Technology. *Paint Coat. Ind.* **2014**, *44*, 31–36, 42.
6. Li, H.J.; Zhang, Z.X.; Qu, H.X.; Ji, L.J.; Xi, J.Q. The Preparation of the Silver Electronic Paste for the Varistors. *Precious Met.* **2017**, *38* (Suppl. 1), 103–107.
7. Liu, X.Q.; Su, X.L.; Liu, X.F.; Qu, Y.H. Preparation and properties of high temperature conductive copper paste. *Electron. Compon. Mater.* **2015**, *34*, 32–35.
8. Ma, X.Q.; Zhu, X.Y.; Long, J.M.; Cao, M. Effect of Glass Powder on Performance of Copper Conductor Film Prepared via Sintering Cu-glass Paste. *Chin. J. Mater. Res.* **2017**, *31*, 472–480.
9. C, F.H.; P, L.C.; S, C.H.; L, H.C. Effects of Fine Particle Content in Al Paste on Screen Printed Contact Formation and Solar Cell Performance. *J. Electrochem. Soc.* **2010**, *157*, H455–H458.
10. Chiu, Y.S.; Cheng, C.L.; Whang, T.J.; Ji, G.Y. Effects of silicate glasses in aluminum pastes on physical and electrical characteristics of screen-printed multi-crystalline silicon solar cells. *Mater. Lett.* **2014**, *126*, 143–146. [CrossRef]
11. Rauer, M.; Schmiga, C.; Raugewitz, A.; Glatthaar, M.; Glunz, S.W. Theoretical and experimental investigation of aluminum-boron codoping of silicon. *Prog. Photovolt. Res. Appl.* **2016**, *24*, 219–228. [CrossRef]
12. Trunov, M.A.; Umbrajkar, S.M.; Schoenitz, M.; Mang, J.T.; Dreizin, E.L. Oxidation and melting of aluminum nanopowders. *J. Phys. Chem. B* **2006**, *110*, 13094–13099. [CrossRef] [PubMed]
13. Trunov, M.A.; Schoenitz, M.; Zhu, X.; Dreizin, E.L. Effect of polymorphic phase transformations in Al_2O_3 film on oxidation kinetics of aluminum powders. *Combust. Flame* **2005**, *140*, 310–318. [CrossRef]
14. Coulet, M.V.; Rufino, B.; Esposito, P.H.; Neisius, T.; Isnar, O.; Denoyel, R. Oxidation mechanism of aluminum nanopowders. *J. Phys. Chem. C* **2015**, *119*, 25063–25070. [CrossRef]
15. Wang, D.X.; Jin, J.S. Analysis of several factors affecting the properties of glass. *Glass* **2004**, *1*, 34–37. (In Chinese)
16. Kaur, K.; Singh, K.J.; Anand, V. Structural properties of Bi_2O_3–B_2O_3–SiO_2–Na_2O glasses for gamma ray shielding applications. *Radiat. Phys. Chem.* **2016**, *120*, 63–72. [CrossRef]

17. Muto, D.; Hashimoto, S.; Daiko, Y.; Honda, S.; Iwamoto, Y. Growth mechanism of house-of-cards aggregates of alumina platelet containing Na_2O-B_2O_3-SiO_2 glass flux. *Ceram. Int.* **2020**, *46*, 9109–9118. [CrossRef]
18. ElBatal, F.H.; Selim, M.S.; Marzouk, S.Y.; Azooz, M.A. UV-vis absorption of the transition metal-doped SiO_2–B_2O_3–Na_2O glasse *Phys. B Condens. Matter* **2007**, *398*, 126–134. [CrossRef]
19. El–Egili, K. Infrared studies of Na_2O–B_2O_3–SiO_2 and Al_2O_3–Na_2O–B_2O_3–SiO_2 glasses. *J. Phys. B.* **2003**, *325*, 340–348. [CrossRe
20. Mei-Yu, C.; Juuti, J.; Chi-Shiung, H.; Chih-Ta, C.; Heli, J. Dielectric $BaTiO_3$-BBSZ glass ceramic composition with ultra-lo sintering temperature. *J. Eur. Ceram. Soc.* **2015**, *35*, 139–144.
21. Siriwardane, R.V.; Poston, J.A.; Robinson, C.; Simonyi, T. Effect of Additives on Decomposition of Sodium Carbonate: Precombu tion CO_2 Capture Sorbent Regeneration. *Energy Fuels* **2011**, *25*, 1284–1293. [CrossRef]
22. Yoshimori, T.; Asano, Y.; Toriumi, Y.; Shiota, T. Investigation on the drying and decomposition of sodium oxalate. *Talanta* **1978**, 2 603–605. [CrossRef]
23. Pujar, P.; Gupta, B.; Sengupta, P.; Gupta, D. Mandal, Sodium ion incorporated alumina—A versatile anisotropic ceramic. *J. Eu Ceram. Soc.* **2019**, *39*, 4473–4486. [CrossRef]

coatings

MDPI

Article

Aluminum Electrodeposition on the Surface of Boron Carbide Ceramics by Use EMIC–$AlCl_3$ Ions Liquid

Roujia Gou [1,*], Jae-Hyeok Park [2], Seiji Yamashita [3], Takeshi Hagio [1,2], Ryoichi Ichino [1,2] and Hideki Kita [1]

[1] Department of Chemical Systems Engineering, Graduate School of Engineering, Nagoya University, Furo-cho, Chikusa-ku, Nagoya 464-8603, Japan
[2] Institute of Materials Innovation, Institutes of Innovation for Future Society, Nagoya University, Furo-cho, Chikusa-ku, Nagoya 464-8601, Japan
[3] Department of Materials Process Engineering, Graduate School of Engineering, Nagoya University, Furo-cho, Chikusa-ku, Nagoya 464-8603, Japan
* Correspondence: gou.roujia@f.mbox.nagoya-u.ac.jp

Abstract: Coating technology is decisively important for metallization of ceramic materials and ceramic metal sealing technology. Previous studies have shown that the network-like structure after penetration of molten aluminum can significantly improve the strength of joint components. However, the direct aluminum coating method is limited by the shape of the substrate. To obtain a dense aluminum film on the surface of B_4C, in this study, aluminum was deposited by pulse electroplating in EMIC–$AlCl_3$ ionic liquid. The deposited metals were observed and analyzed by SEM–EDS and XRD. A Vickers hardness tester was adopted as an auxiliary equipment to clarify the film quality. The results show that frequency and duty cycle have significant effects on crystal orientation. The content of oxides in the contact gap reduces the bonding strength of the deposited metal, which provides experimental basis for metal electrodeposition on B_4C.

Keywords: B_4C ceramics; EMIC-$AlCl_3$; electrodeposition; aluminum

1. Introduction

Boron carbide (B_4C) has the advantages of a high melting point, high hardness, low density, good wear resistance, strong acid and alkali resistance, and high electrical conductivity [1]. Hence, it has been used in various fields including in chemical industries and machinery. However, because of its low fracture toughness, the damage caused during processing and production decreases the material strength to below its ideal value. Moreover, under normal circumstances, the firing temperature of B_4C ceramics is approximately 2200 °C, the energy consumption for firing is high, and it cannot be recycled after being damaged. Therefore, if the fracture toughness of the ceramic material can be improved or its surface can be restored, it can not only extend the service life of ceramics, but also save energy and protect the environment. Metals have excellent electrical and thermal conductivities and ductility and are easy to process. Previous studies [2–6] have shown that a combination of metals and ceramics can yield composite materials with better performance.

Sekine et al. [7] showed that at a temperature of 700 °C, an infiltration phenomenon occurs in the gap between the metal Al and the B_4C joints, and the joint strength is greater than that of the base material itself. Osada et al. [8] also mentioned the use of metal fluidity at high temperatures to heal ceramic materials. In the bonding process, an aluminum (Al) foil is typically used as the bonding material. Although the direct lamination method is simple to operate, this method performs better on parts with simple shapes. For components with complex structures, the bonding performance of the direct coating method is not ideal. The surface metallization treatment of B_4C with metal Al and the method of heating to make the metal penetrate cracks may help cure the surface cracks.

Metallization at the ceramic surface is important for dependable joints of ceramic metals. However, different crystal lattices and types of bonds lead to low wettability. In addition, a significant difference in the coefficient of thermal expansion and melting point temperature makes effective bonding between ceramics and metals difficult.

The possible materials for metallization are limited. Chmielewski et al. [9] proposed a process experiment to prepare Mo–Al_2O_3 composites by hot pressing to overcome brittleness, which is a major technical limitation in the wide application of ceramic materials. Lee et al. [10] used the surface modification technology to braze Al_2O_3–SUS304 with a conventional Ag–Cu eutectic solder. Olesińska et al. [11] described the results of the formation of a barrier layer when copper was attached to copper on AlN ceramics using the copper direct connection technique. The second, more common method, is to create a thin metal coating on the ceramic surface, which can be attached to the metal relatively easily using conventional welding methods. Physical deposition methods, such as plasma spray [12], physical or chemical vapor deposition [13], electroless plating [14], and hot dipping [15], are often used to obtain metal coatings on ceramic surfaces. However, compared with these methods, the electrodeposition method has the advantages of no need heating and better adaptability for complex shape components when deposition, simple operation, and good control of the sedimentary layer structure, and it has been widely used in industries. There has been significant research in recent years, and this method has been improved. Al and its alloys are promising materials because of their low density, high corrosion resistance, and high electrical conductivity. Due to the active chemical properties of Al, the precipitate potential is lower than that of hydrogen ions, and Al cannot be precipitated successfully in aqueous electrolytes. A mixture of 1-ethyl-3-methyl-imidazolium chloride (EMIC) and aluminum chloride ($AlCl_3$) has been used as an ionic liquid for the electrodeposition of Al, using with various metals and alloys, such as Al [16], Al–Zn [17], Al–Mo–Ti [18], and even AlInSb semiconductors [19], have been electrodeposited.

Pulsed current technology has been used in the electrodeposition of Al [20,21] because it can be controlled and improved by adjusting the pulse parameters such as the frequency, duty cycle, pulse strength, and polarity. Yang et al. [20] studied the effects of electrolyte composition, current density, and current form on the characteristics of the Al layer deposited by the $AlCl_3$-n-butylpyridine chloride molten salt system. The authors found that the Al layer deposited by the pulsed current method was denser and smoother than that deposited using the direct current method. Li et al. [21] investigated the influence of pulse parameters on the deposition of Al by an $AlCl_3$–EMIC ionic liquid containing $NdCl_3$. They showed that the micro-deposition changed from polygonal crystals to spherical crystals with increasing DC current and that the pulsed current made the micro-deposition brighter and flatter than that under the DC method at the same average current density.

Kita et al. [7,22] showed that the penetration of pure Al film with the direct coating method into cracks improves the mechanical properties of the material. In our previous study [23], the method employed by Kita et al. was used to obtain coatings; however, it was limited because of some complicated parts. Electroplating, by contrast, is shape-free and easier to control. Although B_4C is known to exhibit conductivity, no study has described the formation of an Al film by electroplating using these properties. Therefore, based on industrial requirements, we investigated electrodeposition coating and deposited Al on B_4C using the EMIC–$AlCl_3$ ionic liquid.

The purpose of this study was to determine the optimum process conditions for obtaining a dense film of Al on the surface of B_4C by electroplating and to clarify the film quality under varying conditions.

2. Materials and Methods

2.1. Basic Parameters of B_4C

B_4C (provided by Mino Ceramics Co., Ltd., Tokyo, Janpan) was used as the substrate. Substrate samples were cut into cuboids with dimensions of 3 mm × 4 mm × 7 mm and mechanically polished in air. Subsequently, they were ultrasonically degreased in ethanol and acetone for 15 min. Table 1 lists its basic parameters.

Table 1. Basic parameters of B_4C tested.

Density (g/cm^3)	Young's Modulus (GPa)	Four-Point Bending Stress (MPa)	Electrical Conductivity (at 25 °C) (S/m)
2.45	430	240 ± 14	140 [24]

2.2. Ionic Liquid Preparation and Electrodeposition

An acidic EMIC–$AlCl_3$ ionic liquid was prepared using EMIC (97%, Tokyo Chemical Industry Co., Ltd., Tokyo, Japan) and $AlCl_3$ (99.99%, Kojundo Chemical Lab. Co., Ltd., Sakado, Japan) in a glass beaker and mixed with a molar ratio of 1:2 at room temperature. The mixture was stirred for over 1 h to ensure homogeneity. An Al wire (99.99%, The Nilaco Corporation, Tokyo, Japan) was soaked in prepared EMIC–$AlCl_3$ ionic liquid for more than one week to remove trace amounts of water remaining in the ionic liquid.

B_4C substrates were priorly rinsed by ultrasonication in distilled water, 0.1 M sodium hydroxide solution, and acetone, respectively, to remove surface impurities, such as grease. In the electroplating experiment, a pure Al wire was used as the anode and Al rod was used as the reference electrode. The electrodeposition area of B_4C was fixed at 2.0×0.3 cm^2, and the extra area was masked with insulation tape. The ambient temperature was 25 °C. Figure 1 shows a schematic diagram of the Al electroplating process using EMIC–$AlCl_3$ ionic liquid. The relevant chemical reactions are as follows [25,26]:

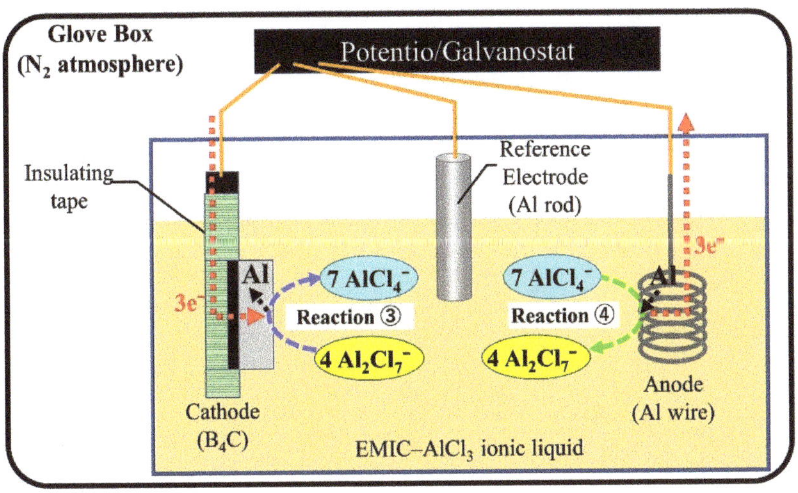

Figure 1. Schematic diagram of the Al electroplating process using EMIC–$AlCl_3$ ionic liquid.

① Ionic liquid: $AlCl_3 + EMIC \rightarrow AlCl_4^- + EMI^+$
② Internal reaction: $AlCl_4^- + AlCl_3 \rightarrow Al_2Cl_7^-$
③ Cathode reaction (reduction): $4Al_2Cl_7^- + 3e^- \rightarrow Al(s) + 7AlCl_4^-$
④ Anode reaction (oxidation): $Al(s) + 7AlCl_4^- \rightarrow 4Al_2Cl_7^- + 3e^-$

Protons have a particularly stronger ability to bind with electrons rather than Al ions; thus, the inclusion of water causes hydrogen generation and inhibits Al deposition reaction in the ionic liquid. Moreover, Al has an active chemical nature, making it easy to react with oxygen to form alumina. To impair the effect of these factors, all experiments were conducted in a glove box filled with N_2. In addition, the uniform coating cannot form on boron carbide ceramics by constant current. It was due to this that the pulse current was used for this experiment. There are three parameters for controlling the deposition, which are current density, frequency, and duty ratio, to obtain the uniform formation. The experiments were conducted to confirm the effects in different variate. Table 2 presents the specific experimental conditions when applying a constant charge of 60 C·cm^{-2} (theorical plating thickness: approximately 20 μm).

Table 2. Experimental conditions.

Sample	Current Density (mA/cm^2)	Frequency (Hz)	Duty Ratio
dAl-1	−15	100	0.8 (8 ms-on/2 ms-off)
dAl-2	−30	100	0.8 (8 ms-on/2 ms-off)
dAl-3	−50	100	0.8 (8 ms-on/2 ms-off)
dAl-4	−30	10	0.8 (80 ms-on/20 ms-off)
dAl-5	−30	50	0.8 (16 ms-on/4 ms-off)
dAl-6	−30	50	0.5 (10 ms-on/10 ms-off)
dAl-7	−30	50	0.2 (4 ms-on/16 ms-off)

The dAl-1~3: effect of current density (−10, −30, −50 mA /cm^2). The surface cover density of coating judged by SEM, then choose the −30 mA/cm^2 as an in-variant current for the next step. The dAl-2, 4, 5: effect of frequency (current density fixed at −30 mA/cm^2, frequency is 100, 10, 50 Hz). Cover density and particle size of coating was judged by SEM observation, then choose the 50 Hz as an invariant frequency for the next step. The dAl-5, 6, 7: effect of duty ratio (fixed current density at −30 mA/cm^2, fixed frequency at 50 Hz, duty ratio 0.8, 0.5, 0.2).

2.3. Analysis Method

The microstructures of the surfaces and cross sections and the elemental analysis of the deposits were observed using a field emission scanning electron microscope (FE-SEM, JEOL, JSM-7500F, Tokyo, Japan). The phase composition was determined by X-ray diffraction (XRD, Rigaku SmartLab, Tokyo, Japan) with Cu Kα radiation (λ = 1.5418 Å). The operating parameters were 40 kV and 30 mA, with a 2θ step size of 0.01. A Vickers hardness tester (FM-300e, FUTURE-TECH CORP, Kanagawa, Japan) was used to apply force and judge the quality of the deposition by changing the surface morphology. In order to reach the substrate, the load for judging the quality of deposition was 5 kg with holding time of 10 s. The load for measurement of the surface hardness by Vickers hardness tester was 0.1 kg.

3. Results

3.1. Observation Results

Figure 2 shows the SEM observation images after electrodeposition. The appearance of the Al film after deposition was silvery white. The Al particle sizes in samples dAl-2, dAl-3, and dAl-5 were smaller than those in the samples dAl-1, dAl-4, dAl-6, and dAl-7, which presented agglomerated and spherical shapes, as shown in the observation results. In addition, the dAl-1, dAl-4, and dAl-7 present poor surface coverage and fewer deposition particle connections. The coating on dAl-7 has a slight peeling, indicating a poor adhesion between the coating and the substrate material.

The dAl-1, dAl-2, and dAl-3 exhibit deposition morphology changes under different current densities (100 Hz, duty ratio of 0.8). When compared with dAl-5, dAl-6, and dAl-7, the SEM observation results show that when the current density and frequency are the same and the duty ratio is 0.5, larger crystals are obtained; however, the Al film deposited is not dense enough.

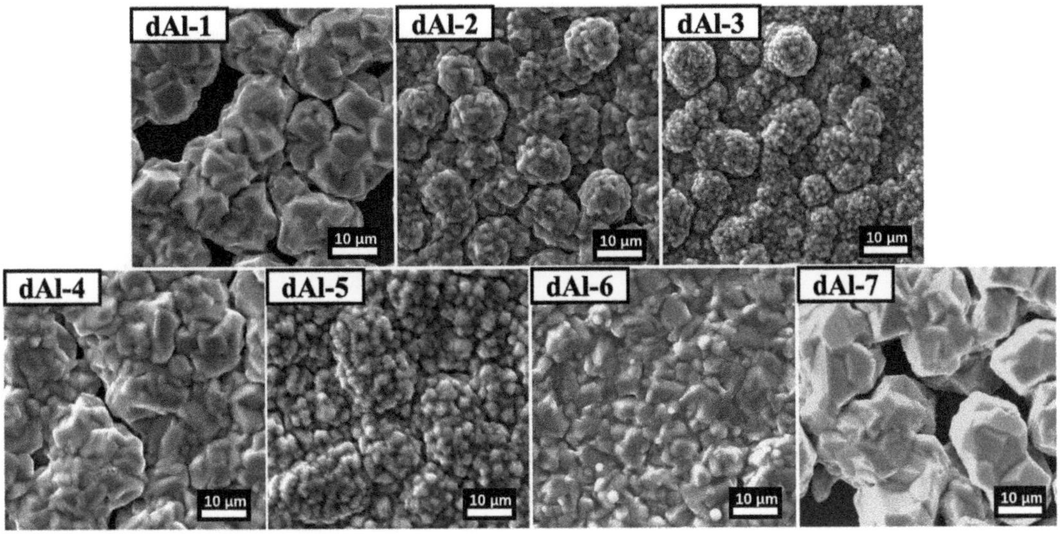

Figure 2. Surface morphologies of Al coatings on B_4C from EMIC–$AlCl_3$ ionic liquid under the different pre-treatments.

3.2. XRD Analysis Results

Figure 3 shows the XRD patterns of each sample. The (111), (200), (220), (311), and (222) surface characteristic peaks of Al can be detected. Metal Al is a face-centered cubic lattice structure which the (111) plane is the close-packed plane. With an increase in the current density, the intensity of the peak (111) decreases. This indicates that at a current density of -15 mA/cm^2, the area of (111) surface was larger. The difference in the diffraction peak heights between dAl-1 and dAl-2 is small, but compared with dAl-3, this difference is significant. The Al peak value of dAl-5 was the highest among the XRD patterns of dAl-2, dAl-4, and dAl-5. With the same plating time, the grain growth increased at a frequency of 10 Hz, which also increase the Al coverage rate.

The peak intensity of (111) for dAl-6 is the same as that of dAl-7. However, the (200) surface had the highest intensity in the dAl-7 sample. Based on the change in the intensity ratios of dAl-2, dAl-4, and dAl-5, the (200) surface intensity increased when the frequency was increased to 50 Hz. Therefore, we suggest that the crystal orientation changes with the frequency.

The crystal size was calculated using the Scherrer equation [27]. The detailed calculation results are presented in Table 3. The average grain size of the Al coatings was calculated from XRD patterns combined with the Scherrer equation given by [28,29]:

$$FWHM = (k \times \lambda)/(L \times cos\ \theta)$$

in which $FWHM$ is the full width at half-maximum of the diffraction peak, k is a Scherrer's constant, L is the crystallite size, and θ is the Bragg angle.

To have the same thickness, the treatment times were different. The thickness of each film was measured by SEM observation and estimates. Compared with dAl-1, dAl-2, and dAl-3, which were deposited under different current densities, in Figure 2 and Table 2, when current densities decreased, the crystal size grew. Since the amount of current affects nucleation and growth, the higher the current density, the faster the nucleation speed and the smaller the particle size. At low current density, the duty ratio was unchanged, and the on-off time was the same, resulting in the growth rate of the crystal being greater than the nucleation rate. This shows that at low current densities, the size of the crystals and the size of the deposited particles grow. In addition, a linear relationship appears when particle size is plotted versus inverse over potential. Furthermore, within each current density stratum, the Al particle size increases as the amount of Al deposited is increased. This effect is most pronounced with low current density deposition due to the relatively low particle density which causes individual particles to grow significantly with increasing Al capacity [25,30]. As shown in Table 3, the grain size of dAl-5 was the smallest with a duty ratio of 0.8. In addition, the growth rate of the surface of the sedimentary layer increases, and the greater spherical protrusions have made the surface rougher. There were nonsignificant effects under frequency changes as shown in the results of dAl-2, dAl-4, and dAl-5, but a high frequency reduced the grain size. With the increase in the frequency, the current duration is shortened, and the surface levelling is improved. However, when the frequency is increased to a certain extent, the interval to change the time-on of the current is too short to supplement metal ions to the cathode, and, at the same time, the hindering effect of the adsorbed substances on the effective interference of grain growth is reduced. Thus, the growth rate is greater than the nucleation rate, and the grain size is increased [31].

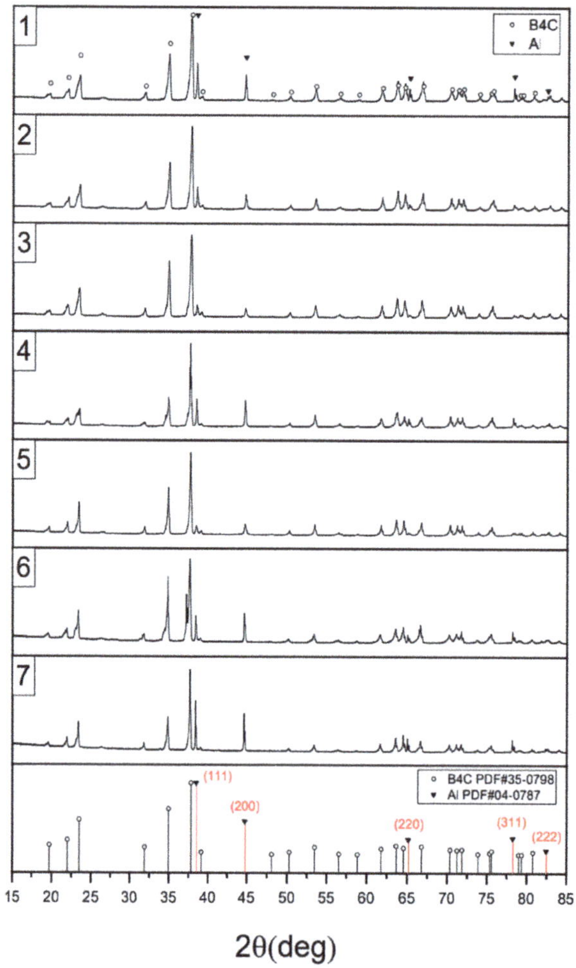

Figure 3. XRD analyses of samples in Table 2.

Table 3. Vickers hardness, plating thickness, and XRD analysis results.

Sample	Hv	Average Size (nm)	Intensity Ratio (111/200)	Thickness (μm)
dAl-1	45.1	63 ± 2	1.49	20 ± 0.28
dAl-2	95.3	39 ± 1	1.55	22 ± 0.78
dAl-3	93.5	35 ± 1	1.41	19 ± 0.56
dAl-4	46.5	56 ± 1	1.01	19 ± 0.37
dAl-5	90.6	34 ± 2	0.89	20 ± 0.25
dAl-6	47.4	59 ± 3	1.11	18 ± 0.91
dAl-7	42.6	62 ± 1	1.47	18 ± 0.80

Based on the observation results of the dAl-5, dAl-6, and dAl-7, the reduced duty cycle and long disconnect time increases the time for grain growth. Therefore, the SEM observation results show that the sizes of the dAl-6 and dAl-7 are larger than those of dAl-5. The calculated grain size results are also consistent with the observed results. Moreover, from the intensity ratios of (111) and (200), it can be found that with an increase in the duty ratio, the preferred orientation of (111) decreases, while the preferred orientation of (200) increases.

3.3. Observation Results of Indentation Treatment

A Vickers hardness tester was used for detecting film bonding condition; the indentation was desired to run through the coating. For this reason, we performed on the coating side with a load of 5 kg and a pressure holding time of 10 s. The indentation did not treat the surfaces of dAl-1, dAl-4, and dAl-7 because of their low surface coverage. From the SEM observation results shown in Figure 4, fractures can be seen near the indentation area in dAl-3 and dAl-5, which are not observed in the other samples. However, the film increased around the indentation area of dAl-2 after loading, which is similar to the results of dAl-3 and dAl-5. The most destroyed face appeared in dAl-5, and the crack was the longest. The complete indentation morphology of sample dAl-6 was obtained, and no ridges or fracture defects were produced. This indirectly proves that the performance of the surface coating meets the requirements.

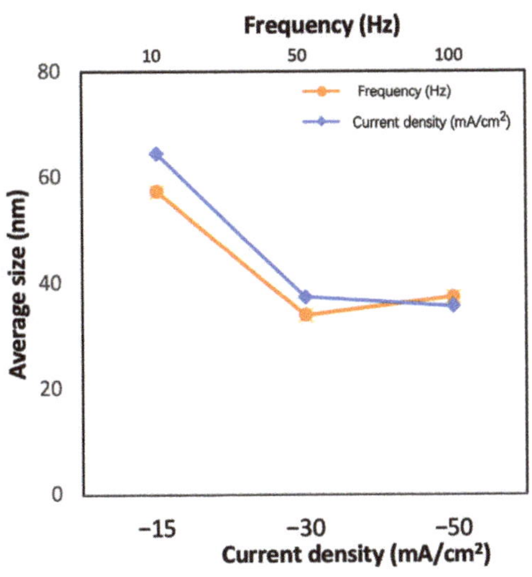

Figure 4. The relation between current density/frequency and an average particle size.

In addition, we compared the indentation position with the deposition surface observation map. As shown in enlarged observation of (c-1) and (d-1), the grains of dAl-5 are clearly separated, and the indentation position can be clearly seen even after plastic deformation under stress. Therefore, compared to dAl-6, dAl-5 has a higher brittle phase content, which affects the toughness strength of the deposited Al and leads to the generation of cracks. From the observation images of the cracks shown in Figure 5, we suggest that an intergranular fracture occurred at the deposited surface after loading. In order to observe the cross-section surface of coating, the three-point bending tester was used to break the sample in two pieces. After testing, the Al film of dAl-6 did not break into two pieces but peeled from the surface of B_4C. Figure 6 shows the cross-section SEM observation after the 3-point bending test. In Figure 6, it can be observed that the (a) and (b) coatings are arching on the surface of B_4C due to the external force; (c) and (d) did not escape from the substrate because the deposition particles gap released the force. In addition, we observed the metal fracture morphology of dAl-2, which had no cracks after the indentation treatment, and dAl-5, which had more surface cracks. The results, presented in Figure 7a–c represent the vertical direction of the observed crack and morphology of dAl-5. Figure 7d,e show the cross-sectional observation results of dAl-5 and dAl-2, respectively. Comparing the EDS mapping images of (d-1) and (e-1) in Figure 7, the oxygen distribution is different. In (e-1), the brittle oxide phase is deposited at the grain boundaries, which prevents the plastic relaxation and has hardening effects. The surface hardness value of smaller grains is higher, and the brittle phase between the grains (such as Al_2O_3) under the action of the external force appeared to have separated and fractured due to plastic deformation pulling, resulting in cracks [32,33].

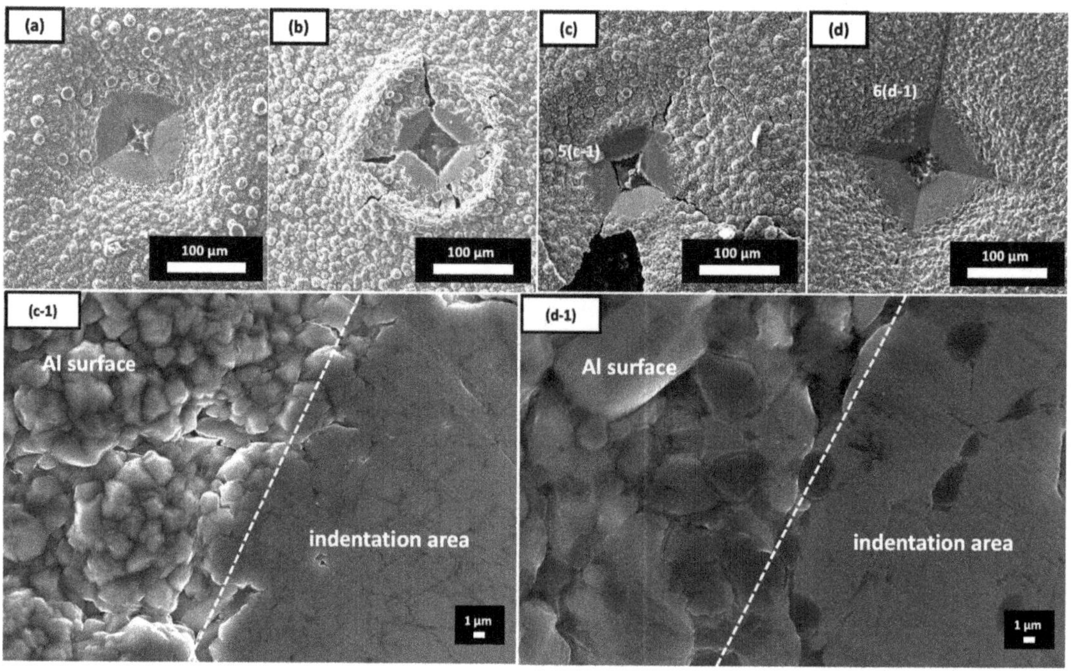

Figure 5. Indentation morphologies observations: (**a**): dAl-2; (**b**): dAl-3; (**c**): dAl-5; (**d**): dAl-6 by SEM; (**c-1**) and (**d-1**) are the enlarged observation of (**c**) and (**d**) separately.

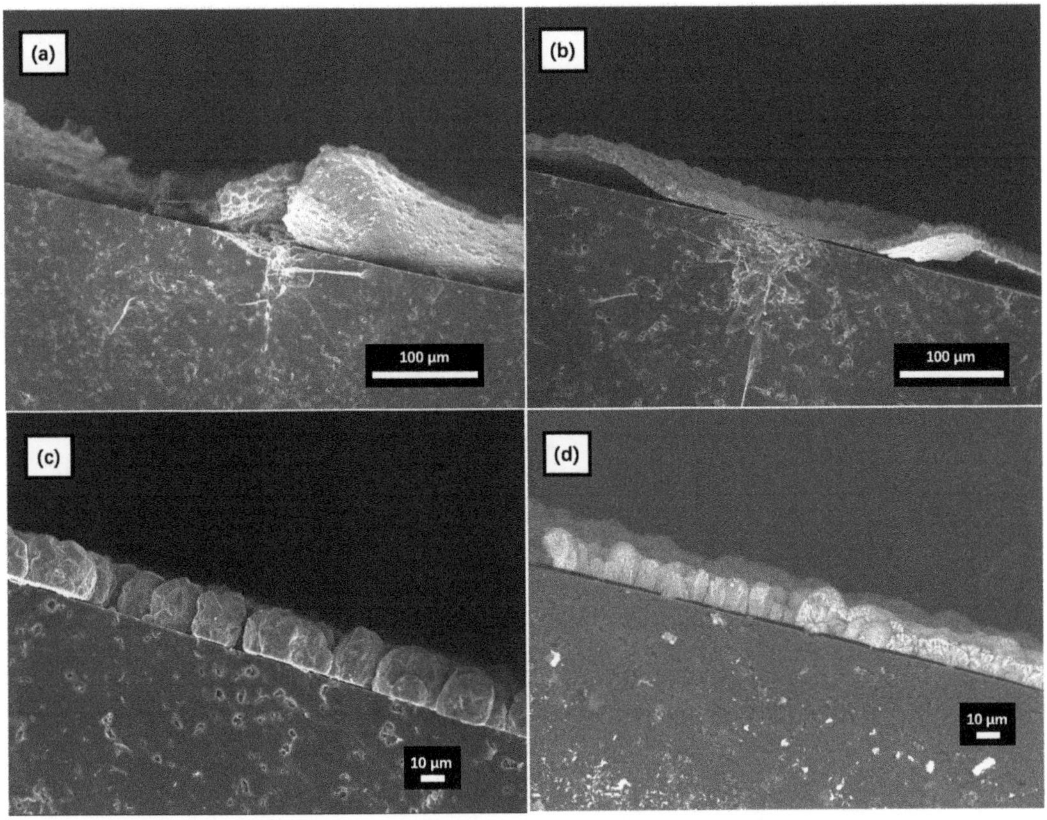

Figure 6. The cross-section SEM observation after 3-point bending test, (**a**) dAl-2; (**b**) dAl-5; (**c**) dAl-7; (**d**) dAl-4.

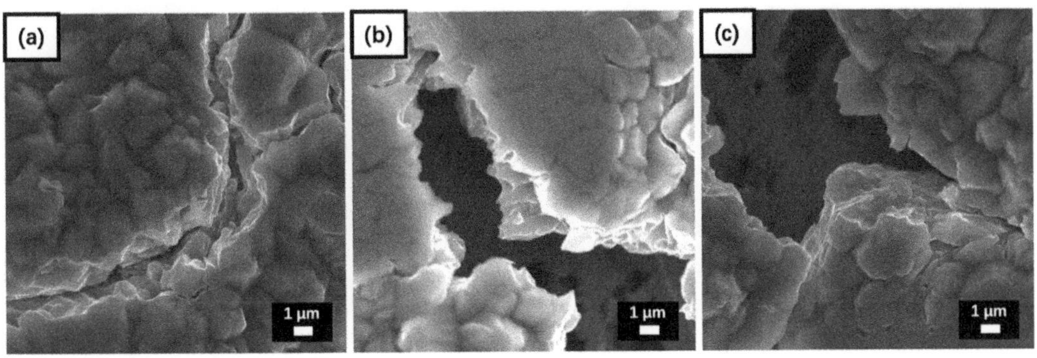

Figure 7. *Cont.*

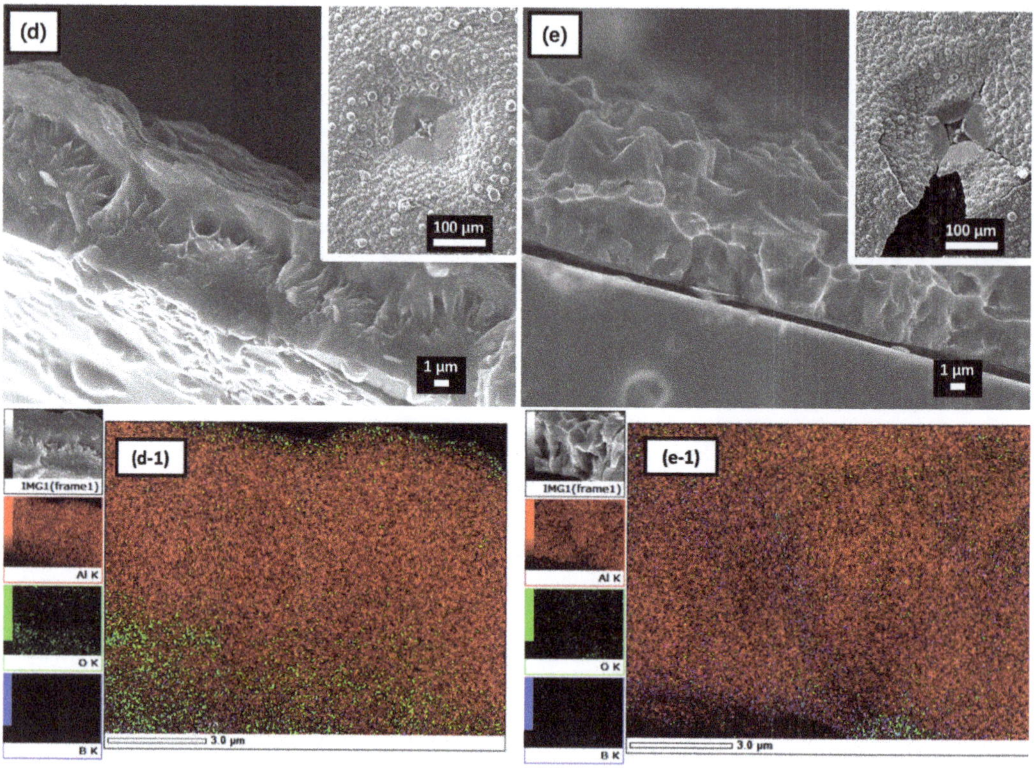

Figure 7. Observation results of cracks in sample dAl-5 are (**a–c**); (**d**,**e**) are cross-sectional observation results of dAl-2 and dAl-5, respectively; (**d-1**) and (**e-1**) EDS mapping images of cross-sectional observation.

Observations of the cracks in Figure 7 show that there are two types of fracture marks: the grain boundary fracture and the deformation tearing. The crystals of the Al on the fractured surface were relatively small and granular. From the observation results that the specific surface area of the deposited Al with a smaller particle size is higher; therefore, the Al film is oxidized in the gap as a brittle phase, which increases the hardness of the deposited Al on the surface. In addition, differences in elemental distribution can be seen by EDS surface scans. The analysis of the samples in combination with the morphology map reveals that the distribution of alumina is more pronounced in the fracture. Indirectly, it has proved that the secondary phase oxide has an influence on the quality of the deposited Al.

4. Conclusions

The grain size decreased with increasing deposition current density. In addition, the enhanced concentration polarization decelerated the deposition rate and decreased the thickness of the deposited film. The alumina film formed in the gap as a brittle phase, which increases the hardness of the deposited Al, but the oxides in the contact gap reduces the bonding strength of the deposited particles.

With the increase in the frequency, the current duration is shortened, and the surface levelling is improved. However, when the frequency is increased to a certain extent, the interval to charge the time-on of current is too short to supplement metal ions to the cathode, and, additionally, the hindering effect of the adsorbed substances on the effective interference of grain growth is reduced. Thus, the growth rate is greater than the nucleation rate, and the grain size is increased. At low frequencies, increasing the duty cycle increased the time for grain growth. The duty ratio had a significant influence on the preferred orientation. Under a duty ratio of 0.5, the preferred orientation level of (200) was higher than that of (111).

In this study, a Vickers hardness tester was used to apply force and judge the quality of the deposition by changing the surface morphology. The dAl-6 (−30 mA, 50 Hz, 0.5) indentation observation shows no cracking appearance, which indirectly reflects the great bonding strength of deposition particles. The Vickers hardness tester practicable stands a chance of being a useful adjunct to judge the quality of the deposited film.

Author Contributions: Conceptualization, R.G. and H.K.; methodology, J.-H.P., T.H., R.I., R.G., S.Y. and H.K.; validation, T.H., S.Y. and R.G.; formal analysis, J.-H.P. and R.G.; investigation, J.-H.P., T.H., R.G. and S.Y.; resources, R.I. and H.K.; writing—original draft preparation, R.G.; writing—review and editing, J.-H.P., T.H., R.I., H.K., S.Y. and R.G.; supervision, R.I. and H.K.; funding acquisition, H.K. All authors have read and agreed to the published version of the manuscript.

Funding: This work has been financially supported by Japan Science and Technology Agency (JST)—Program on Open Innovation Platform with Enterprises, Research Institute and Academia (OPERA), Grant JPMJOP1843.

Institutional Review Board Statement: Not applicable.

Informed Consent Statement: Not applicable.

Data Availability Statement: Data is contained within the article.

Conflicts of Interest: The authors declare no conflict of interest.

References

1. Thévenot, F. Boron carbide—A comprehensive review. *J. Eur. Ceram. Soc.* **1990**, *6*, 205–225. [CrossRef]
2. Ma, G.; Xiao, H.; Ye, J.; He, Y. Research status and development of magnesium matrix composites. *Mater. Sci. Technol.* **2020**, *36*, 645–653. [CrossRef]
3. Hayat, M.D.; Singh, H.; He, Z.; Cao, P. Titanium metal matrix composites: An overview. *Compos. Part A Appl. Sci. Manuf.* **2019**, *121*, 418–438. [CrossRef]
4. Kumar, V.M.; Venkatesh, C.V. A comprehensive review on material selection, processing, characterization and applications of aluminium metal matrix composites. *Mater. Res. Express* **2019**, *6*, 072011. [CrossRef]
5. Wang, J.; McDannald, A.; Karandikar, P.; Aghajanian, M. Fabrication of reaction bonded TiB 2/Si/SiC composites for thermal applications. *Int. J. Ceram. Eng. Sci.* **2020**, *2*, 264–270. [CrossRef]
6. Bajakke, P.A.; Malik, V.R.; Deshpande, A.S. Particulate metal matrix composites and their fabrication via friction stir processing–A review. *Mater. Manuf. Process* **2019**, *34*, 833–881. [CrossRef]
7. Sekine, K.; Kumazawa, T.; Tian, W.; Hyuga, H.; Kita, H. Low-temperature joining of boron carbide ceramics. *J. Ceram. Soc. Jpn.* **2012**, *120*, 207–210. [CrossRef]
8. Osada, T.; Kamoda, K.; Mitome, M.; Hara, T.; Abe, T.; Tamagawa, Y.; Nakao, W.; Ohmura, T. A Novel design approach for self-crack-healing structural ceramics with 3D networks of healing activator. *Sci. Rep.* **2017**, *7*, 717853. [CrossRef]
9. Chmielewski, M.; Dutkiewicz, J.; Kaliński, D.; Lityńska Dobrzyńska, L.; Pietrzak, K.; Strojny Nędza, A. Microstructure and properties of hot-pressed molybdenum-alumina composites. *Arch. Metall. Mater.* **2012**, *57*, 687–693. [CrossRef]
10. Lee, K.-Y.; Hwang, W.-K.; Jung, I.-S. Brazing joining of Al_2O_3-SUS304 with surface modification method. In Proceedings of the 3rd International Brazing and Soldering Conference (ASM International), San Antonio, TX, USA, 24–26 April 2006; Volume 4, pp. 24–26.
11. Olesin'ska, W.; Kalin'ski, D.; Chmielewski, M.; Diduszko, R.; Wlosin'ski, W. Influence of titanium on the formation of a "barrier" layer during joining an AlN ceramic with copper by the CDB technique. *J. Mater. Sci. Mater. Electron.* **2006**, *7*, 781–788. [CrossRef]
12. Sarikaya, O.; Anik, S.; Aslanlar, S.; Cem Okumus, S.; Celik, E. Al–Si/B_4C composite coatings on Al–Si substrate by plasma spray technique. *Mater. Des.* **2007**, *28*, 2443–2449. [CrossRef]
13. Yun, J.-H.; Kim, B.-Y.; Rhee, S.W. Metal–organic chemical vapor deposition of aluminum from dimethylethylamine alane. *Thin Solid Films* **1998**, *312*, 259–264. [CrossRef]
14. Koura, N.; Nagase, H.; Sato, A.; Kumakura, S.; Takeuchi, K.; Ui, K.; Tsuda, T. Electroless Plating of Aluminum from a Room-Temperature Ionic Liquid Electrolyte. *J. Electrochem. Soc.* **2008**, *155*, D155. [CrossRef]
15. Cheng, W.-J.; Wang, C. Growth of intermetallic layer in the aluminide mild steel during hot dipping. *Surf. Coat. Technol.* **2009**, *204*, 824–828. [CrossRef]

16. Zhao, Y.; VanderNoot, T.J. Electrodeposition of aluminum from room temperature $AlCl_3$–TMPAC molten salts. *Electrochim. Ac* **1997**, *42*, 1639. [CrossRef]
17. Pan, S.-J.; Tsai, W.-T.; Chang, J.-K.; Sun, I.W. Co-deposition of Al–Zn on AZ91D magnesium alloy in $AlCl_3$–1-ethyl- methylimidazolium chloride ionic liquid. *Electrochim. Acta.* **2010**, *55*, 2158–2162. [CrossRef]
18. Tsuda, T.; Arimoto, S.; Kuwabata, S.; Hussey, C.L. Electrodeposition of Al–Mo–Ti Ternary Alloys in the Lewis Acidic Aluminur Chloride–1-Ethyl-3 methylimidazolium Chloride Room Temperature Ionic Liquid. *J. Electrochem. Soc.* **2008**, *155*, D256. [CrossRe
19. Tsuda, T.; Hussey, C.L. Electrodeposition of photocatalytic AlInSb semiconductor alloys in the Lewis acidic aluminum chloride ethyl-3-methylimidazolium chloride room-temperature ionic liquid. *Thin Solid Films* **2008**, *516*, 6220–6225. [CrossRef]
20. Yang, C. Electrodeposition of aluminum in molten $AlCl_3$-n-butylpyridinium chloride electrolyte. *Mater. Chem. Phys.* **199** *37*, 355–361.
21. Li, B.; Chen, Y.; Yan, L.; Ma, J. Pulse current electrodeposition of Al from an $AlCl_3$-EMIC ionic liquid containing NdCl *Electrochemistry* **2010**, *78*, 523–525. [CrossRef]
22. Sekine, K.; Kumazawa, T.; Tian, W.; Hyuga, H.; Kita, H. Influence of joining time and temperature on the flexural strength c joined boron carbide ceramics. *J. Ceram. Soc. Jpn.* **2012**, *120*, 393–399. [CrossRef]
23. Gou, R.; Yamashita, S.; Sekine, K.; Kita, H. Microstructural analysis of network-like crack structure formed at Al–B_4C interface. *Eur. Ceram. Soc.* **2021**, *41*, 6962–6970. [CrossRef]
24. Tan, Y.; Zhang, H.; Peng, S. Electrically conductive graphene nanoplatelet/boron carbide composites with high hardness an toughness. *Scr. Mater.* **2016**, *114*, 98–102. [CrossRef]
25. Zhang, M.; Kamavaram, V.; Reddy, R.G. New electrolytes for aluminum production: Ionic liquids. *JOM* **2003**, *55*, 54. [CrossRe
26. Jiang, T.; Chollier Brym, M.J.; Dube, G.; Lasia, A.; Brisard, G.M. Electrodeposition of aluminum from ionic liquids: Part I- Electrodeposition and surface morphology of aluminum from aluminum chloride ($AlCl_3$)–1-ethyl-3-methylimidazolium chlorid ([EMIm]Cl) ionic liquids. *Surf. Coat. Technol.* **2006**, *201*, 1–9. [CrossRef]
27. Holder, C.F.; Schaak, R.E. Tutorial on powder X-ray diffraction for characterizing nanoscale materials. *J. Electrochem. Soc.* **198** *136*, 7359–7365. [CrossRef]
28. Scherrer, P. Bestimmung der Grösse und der inneren Struktur von Kolloidteilchen mittels Röntgenstrahlen. *Nachr. Ges. Wis Göttingen* **1918**, *26*, 98–100.
29. Langford, J.I.; Wilson, A.J.C. Scherrer after Sixty Years: A Survey and Some New Results in the Determination of Crystallite Siz *J. Appl. Cryst.* **1978**, *11*, 102–113. [CrossRef]
30. Pei, A.; Zheng, G.; Shi, F.; Li, Y.; Cui, Y. Nanoscale Nucleation and Growth of Electrodeposited Lithium. *Metal. Nano Lett.* **2017**, *1* 1132–1139. [CrossRef]
31. Fukunaka, Y.; Yamamoto, T.Y.; Kondo, T. Electrodeposition of silver under direct and pulsed current. *ACS Nano.* **2019**, *1* 3278–3283. [CrossRef]
32. Gao, Y.; Wang, C.; Yao, M.; Liu, H. The resistance to wear and corrosion of lasercladding Al_2O_3 ceramic coating on Mg alloy. *App Surf. Sci.* **2007**, *253*, 5306–5311. [CrossRef]
33. Luo, X.; Yao, Z.; Zhang, P.; Gu, D. Al_2O_3 nanoparticles reinforced Fe-Al laser cladding coatings with enhanced mechanica properties. *J. Alloy. Compd.* **2018**, *755*, 41–54. [CrossRef]

coatings

MDPI

Article

Evaluation of Fracture Toughness of Plasma Electrolytic Oxidized Al_2O_3-ZrO_2 Coatings Utilizing Nano-Scratch Technique

Mehri Hashemzadeh [1,2,*], Frank Simchen [1], Lisa Winter [1] and Thomas Lampke [1]

[1] Materials and Surface Engineering Group, Institute of Materials Science and Engineering, Chemnitz University of Technology, 09107 Chemnitz, Germany

[2] Gruppenleiter Elektrochemie, INNOVENT e.V. Technologieentwicklung, Prüssingstraße 27, 07745 Jena, Germany

* Correspondence: mehri.hashemzadeh@mb.tu-chemnitz.de

Abstract: Al_2O_3 coatings, which can be produced by plasma electrolytic oxidation (PEO) on aluminum substrates, provide an excellent protection against corrosion and wear. However, due to the brittle nature of the oxide ceramic, the fracture toughness is limited. One approach to enhance the tolerance to fracture is the incorporation of ZrO_2 to form zirconia toughened alumina (ZTA). In addition to its use as a bulk material, the application as a coating material enables a broader field of application. In this study, an Al_2O_3-ZrO_2 composite coating was applied on a 6082 aluminum alloy using an aluminate-phosphate-based electrolytic solution containing a Zr-based salt. Polarization measurement as an indicator of the passivability of a given system revealed that Zr-based salt improves the passivation of the aluminum alloy. The coatings' characteristics were evaluated by SEM, EDS, and XRD. ZrO_2 incorporated into alumina as a metastable high-temperature modification led to a thicker coating with new morphologies including lamellar and dendritic structures. Nano-indentation showed that the incorporated Zr increase the average hardness of the compact layer from 16 GPa to 18 GPa. The fracture toughness of the coatings was investigated locally with nano-scratches applied on the compact outer layer of the coatings' cross-sections. The Zr-containing electrolytic solution resulted in a coating with significantly higher fracture toughness (6.9 $MPa \cdot m^{1/2}$) in comparison with the Zr-free electrolytic solution (4.6 $MPa \cdot m^{1/2}$). Therefore, it is shown, that the PEO process stabilized a high-temperature allotrope of zirconia at room temperature without the need for rare-earth dopants such as Y_2O_3. Furthermore, it was demonstrated that the nano-scratch method is a suitable and accurate technique for the investigation of the fracture toughness of coatings with inherent cracks

Keywords: Al_2O_3-ZrO_2 coating; plasma electrolytic oxidation; fracture toughness; nano-scratch; nano-indentation

1. Introduction

Al_2O_3 is widely utilized in various applications, which require high wear resistance at elevated temperatures, thanks to its excellent mechanical and electric properties. However, Al_2O_3 has deficient fracture toughness and high brittleness [1], which can be a limiting factor for the application. To compensate this disadvantage and improve the fracture behavior, ZrO_2 can be incorporated into Al_2O_3 to form zirconia toughened alumina (ZTA), which is well-known for its high-temperature mechanical strength, appropriate thermal shock resistance and low thermal conductivity as well as high fracture toughness [1,2]. In addition to its use as a bulk material, the application of ZTA as a coating enables a broader field of application due to the combination of a high-temperature and wear-resistant coating and a ductile substrate. Al_2O_3-ZrO_2 composite coatings have been fabricated by thermal spraying [1,3], laser deposition [4], and plasma electrolytic oxidation [2,5–7].

Plasma electrolytic oxidation (PEO) is an advanced form of the anodizing process, which is often implemented in an environmentally friendly alkaline electrolyte under different electrical regimes including direct or alternating current at voltages higher than the ignition voltage leading to micro-discharges across the grown oxide [8–13].

Literature review shows that PEO Al_2O_3-ZrO_2 coating may be produced through addition of ZrO_2 nanoparticles [2], or Zr-based salt additives such as potassium zirconium carbonate ($K_2[Zr(CO_3)_2(OH)_2]$) [14], and dipotassium hexafluorozirconate (K_2ZrF_6) [7,15] into the electrolytic solution. Matykina et al. [2] prepared this coating in phosphate-based and silicate-based electrolytes, which contain monoclinic ZrO_2 nano particles (m-ZrO_2), and observed that the phosphate-based electrolytic solution developed high-temperature forms of zirconia including tetragonal ZrO_2 (t-ZrO_2) and/or cubic ZrO_2 (c-ZrO_2). Both phases could not be distinguished clearly by X-ray diffraction (XRD) because of overlapping diffraction peaks. However, the electrolyte with ZrO_2 nano-particles is a suspension and therefore, it is potentially unstable. Especially in the context of the complex hydrodynamic conditions present within the discharge zone, reproducible coating results can only be expected to a limited extent. This may prove to be a disadvantage during the upscaling of the PEO process.

Rehman et al. [5] prepared corrosion-resistant Al_2O_3-ZrO_2 coatings in a phosphate-based electrolyte through a two-step PEO process using K_2ZrF_6 additive. However, poisonous HF gas may evolve during the PEO process owing to the local acidification of the oxide/electrolyte interface. It was established that $K_2[Zr(CO_3)_2(OH)_2]$ can produce viable ZrO_2-containing coatings trough a PEO process [14,16].

When determining mechanical properties of thin coatings, standardized methods used for bulk materials have to be adapted in most cases. The standardized test method for the determination of fracture toughness of advanced ceramics is defined in ASTM C1421-18 [17]. In addition to this testing method, other strategies based on micro- or nano-indentation with a sharp probe have been established, mostly for analyzing bulk dental ceramics [18]; however, the application for coatings is also possible [19,20]. The common method is the Vickers-indentation fracture test where the fracture toughness, K_c, is calculated by a Vickers probe by using the following equation [21]:

$$K_c = \alpha \left[\frac{E}{H}\right]^{\frac{1}{2}} \left[\frac{F}{c_0^{\frac{1}{2}}}\right] \tag{1}$$

where E and H are Young's modulus and hardness, respectively. F and c_0 are the indentation load and the average length of the radial cracks promoted by the indentation, respectively. α is a dimensionless constant [21]. Equation (1) has undergone numerous modifications in the past few years of research, incorporating factors such as crack type, residual stress, and plastic dissipation within the material. These parameters are obtained using dimensional analysis and empirical data; however, to date, there is no analytical model that fully supports them. In addition, it is critical to measure the length of the crack growing from the four corners of the indent accurately [22]. However, the application of this method for PEO coatings is challenging as the determination of cracks generated by nano-indentation is impeded by the inherent micro-cracks in the coating [23].

Another technique to evaluate the fracture toughness of a wide range of materials including ceramics, metals, and polymers is the nano-scratch test [22]. Four scratch-based methods have been utilized to calculate the fracture toughness: one is derived from linear-elastic fracture mechanics (LEFM), and the other three methods are developed on microscopic energetic size effect laws (MESEL) [24]. It is well-established that LEFM is applicable to brittle materials such as ceramics and glasses [24]. Using LEFM, the following equation was derived (lateral force F_l, the perimeter p, the horizontal projected load bearing contact area A) [25]:

$$K_c = \frac{F_l}{[2pA]^{\frac{1}{2}}} \tag{2}$$

As seen in Equation (2), fracture toughness can be estimated straightforward using nano-scratch tests. For scratch tests, it is necessary to apply normal loads which are neither too large nor too small in order to avoid crack-free and severe-damage regimes [24].

The main focus for the present study is the formation of Al_2O_3-ZrO_2 coatings using PEO and the evaluation of the role of incorporated zirconia phases on the fracture toughness. The PEO Al_2O_3-ZrO_2 coating was prepared using a stable aluminate-phosphate electrolytic solution, which contains $K_2[Zr(CO_3)_2(OH)_2]$ as a non-toxic Zr-based salt. To establish, if zirconia incorporated in the PEO coating could enhance the fracture toughness as it does in bulk materials, nano-scratch tests are utilized for the determination of local fracture toughness.

2. Materials and Methods

2.1. Sample Preparation

Precision-turned cylindrical specimens of a 6082 Al alloy in peak-aged condition with a diameter of 16.5 mm and a height of 10 mm were used as substrates. Heat-shrinkable epoxy tubes were used for masking the specimen to ensure electrical insulation and to prevent any edge effects. The procedure was described in detail in [26]. The substrates were degreased in ethanol, etched in 3 wt%-NaOH solution at 50 °C for 20 s, and finally washed in distilled water and dried.

2.2. Polarization Test

Polarization method is an appropriate technique for the evaluation of the passivation capability of the electrolyte/substrate combination [26]. Therefore, polarization measurements were used for the study of the effect of Zr-based salt on the passivation of the substrate. Polarization measurements were conducted with an applied potential of up to 100 V, which does not trigger discharges. The time-dependent potential $\varphi(t)$ and current density $i(t)$ were recorded during the test. In order to quantify the passivation behavior of the different substrate/electrolyte combinations, the passivation parameter σ [27] was determined for all tests by integrating the $i(t)$ curve between t = 100 and 220 s. A three-electrode system including Ag/AgCl (3 mol/L KCl) as the reference electrode, a platinum sheet as the counter electrode and 6082 Al alloy as the working electrode were used. Further details regarding the method can be found [27].

2.3. Plasma Electrolytic Oxidation Treatment

A power source pe861UA-500-10-24-S (plating electronics, Sexau, Germany) and a capsuled treatment station (Scheigenpflug, Leipzig, Germany) with a 5 L container were used for the production of PEO coatings. The experimental setup including the equipment for process diagnostic is described in detail in [26]. The reference coating was prepared in an electrolytic solution containing 0.2 mol/L $NaAl_2O$ (VWR Chemicals, Radnor, PA, USA) and 0.035 mol/L Na_2HPO_4 (Sigma-Aldrich, Saint Louis, MO, USA). The pH of the bath was adjusted up to 12 using dilute KOH (Sigma-Aldrich, Saint Louis, USA). In order to prepare the Al_2O_3-ZrO_2 composite, 0.1 mol/L of the Zr-based salt $K_2[Zr(CO_3)_2(OH)_2]$ (Lehvoss, Hamburg, Germany) was added to the reference electrolyte and the pH was raised to 12 by KOH. These electrolytes are summarized in Table 1.

Table 1. Composition and pH of the electrolytic solutions used for the production of PEO coatings.

Coating	c $NaAl_2O$ (mol/L)	c Na_2HPO_4 (mol/L)	c $K_2[Zr(CO_3)_2(OH)_2]$ (mol/L)	pH
Reference	0.2	0.035	-	12
Zr electrolyte	0.2	0.035	0.1	12

The process voltage was recorded with a ScopeCorder DL850 (Yokogawa, Musashino, Japan). In addition, the current signal was obtained by measuring the voltage drop over a shunt resistor (Isabellenhütte, Dillenburg, Germany). In order to capture optical process information and electrical process data simultaneously, a photodiode was placed near the working electrode. Further details are described in [26].

2.4. Coating Characterization

To analyze the cross-sectional morphology of the prepared coatings, a scanning electron microscope (SEM, model LEO 1455VP) and a field-emission SEM (FESEM, NEON 40EsB, Zeiss, Jena, Germany) were used. The specimens were cut and embedded in epoxy resin. The cross-sections were ground using SiC papers and polished with diamond paste. Cleaning was carried out ultrasonically in ethanol. Finally, the specimens were dried through airflow. Elemental analyses of the cross-section of the coatings were carried out using Energy Dispersive X-ray (EDX) spectroscopy. The average thickness of the coating was determined using at least 15 measuring points in the cross-section images through the software Image J 1.44p (version 1.6).

The X-ray diffractometer D8 Discover (Bruker, Billerica, MA, USA) was carried out using a Co K_α radiation source with point focus, a 2 mm pinhole aperture, and a LYNXEYE-XE-T detector. Accelerating voltage and tube current were 35 kV and 40 mA, respectively. PDF2 (2014) database of the International Centre for Diffraction data (ICDD) was employed for the identification of the phases.

2.5. Nanoindentation Measurement

A nano-indenter UNAT (ASMEC, Dresden, Germany) with a Berkovich indenter was used for the nanoindentation experiments on the polished cross-section of the coatings. The outer layer of the coating was used, as it was compact and therefore more suitable for subsequent investigations on the fracture toughness. At least 15 indents for each coating were performed using a normal load of 100 mN with a loading time of 10 s, a holding time of 5 s, and an unloading time of 4 s. The Oliver–Pharr method was used for the calculation of the hardness and indentation modulus [28].

2.6. Nano-Scratch Experiment and Methods for Calculating the Fracture Toughness

Nano-scratch tests were carried out with the nano-indenter UNAT (ASMEC, Dresden, Germany) as short, high-resolution scratches using a lateral force unit. The procedure for a standard scratch test includes three main steps: pre-scan, scratching and post-scan which is explained in detail in [29]. Using a spherical diamond tip with a diameter (R) of 10 µm, a linearly progressive normal load (F_n) of up to 300 mN was applied on the polished cross-section of the PEO coatings. The nano-scratch was applied through scanning at 3 µm/s over a 10 µm track. The load of pre- and post-scans was 1 mN. As there is no literature to compare with, the scratching parameters for the PEO coatings were experimentally determined. It is important to mention that the PEO coating without incorporated zirconia served as the reference coating and its lower fracture resistance determined the used nano-scratch parameters. The software of UNAT-InspectorX recorded the variations of penetration depth (P_d), lateral force (F_l), and normal force (F_n) in relation to the scratch distance. To avoid the effect of pores and cracks in the coating on the calculated fracture toughness, the scratches were applied merely on the compact parts of the coating outer layer, which is illustrated in Section 3.3. In addition, it will be shown that Zr affects the microstructure of the outer layer of the PEO coating, significantly. Therefore, the capability of Zr for enhancing fracture toughness of PEO alumina was studied by focusing on the compact outer layer. As the application of nano-scratch tests is a novel strategy for the evaluation of fracture toughness of PEO coatings, they were carried out on three samples of each coating condition for sufficient statistical reliability. Overall, at least 15 nano-scratches were applied for each condition.

For a spherical indenter, Equation (3) can be extracted from Equation (2) [25]:

$$F_l = 4 \cdot K_c \cdot d_p \cdot \sqrt{\frac{2 \cdot R}{3}} \text{ for } d_p << R \quad (3)$$

When using an indenter diameter of $R = 10$ μm, Equation (3) can be rewritten as the following numerical-value equation:

$$F_l = 103.3 \cdot \frac{K_c \cdot d_p}{R} \quad (4)$$

where the units for F_l, K_c and d_p are mN, MPa·m$^{1/2}$, and μm, respectively. Based on Equation (4), K_c can be calculated through the linear fitting of the F_l–d_p/R curve.

3. Results

3.1. Polarization Measurements

Polarization measurements were conducted in the reference electrolyte and the Zr electrolyte. The results are presented in Figure 1.

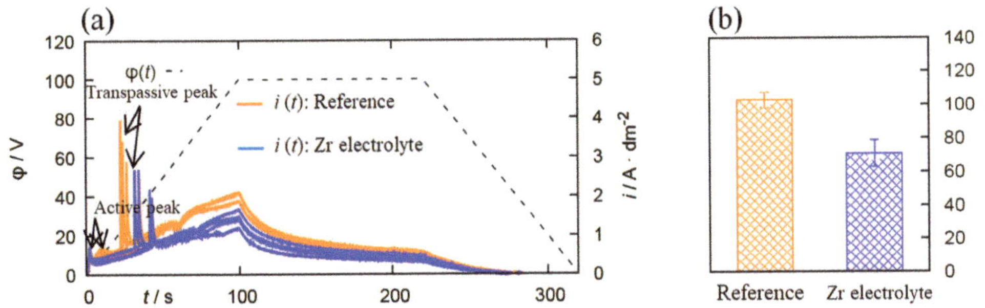

Figure 1. (**a**) $\varphi(t)$- and $i(t)$- diagrams of the polarization measurements recorded on 6082 Al alloy in the reference electrolyte and the Zr electrolyte. (**b**) The passivation parameter σ determined by integrating the $i(t)$ curve between t = 100 and 220 s. With the addition of Zr-based salt into the electrolyte, the passive region of the substrate is extended, which indicates a positive effect regarding the passivation behavior.

The $i(t)$ curve can be divided into the three regions "active", "passive", and "transpassive" based on a model proposed by Krysmann and Kurze [10]. First, an ion-conducting oxide layer is formed on the substrate by the anodic polarization and results in a sharp peak called "active peak" in the $i(t)$ curve. The oxide layer thickens and this appears as a low-slope region indicating a "passive region" in the plot. Continuously increasing voltage allows for the evolution of oxygen molecules on the anode, leading to a sharp increase in the current density ("transpassive region"). As seen in Figure 1a, the addition of Zr-based salt into the bath extends the passive region of the substrate, indicating the positive effect on the passivability of the substrate. This is confirmed by the σ-value, which is (103 ± 5) As/dm^2 for the reference electrolyte and (71 ± 8) As/dm^2 for the Zr containing solution. Lower σ-values correspond to a better passivation.

3.2. Process Diagnostics

The results of the electrical and optical process diagnostics are summarized in Figure 2. Figure 2a shows the course of the anodic and cathodic peak voltage over the process time.

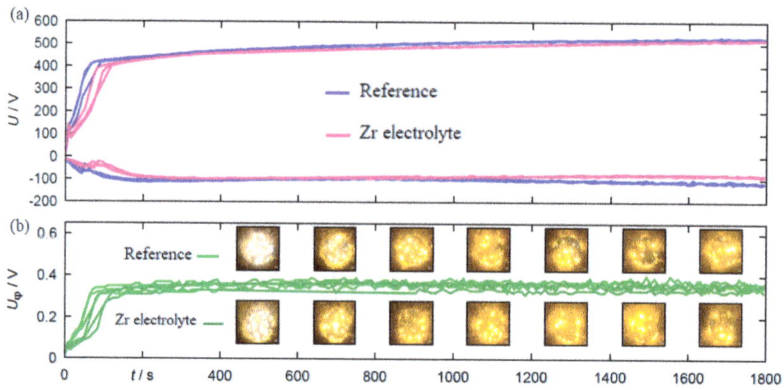

Figure 2. (**a**) The course of anodic and cathodic peak voltage and (**b**) the photo-voltage for specimens coated by PEO process in the reference electrolyte and the Zr electrolyte.

Figure 2b shows the maximum photovoltage as well as single frames of the video recordings placed to the corresponding process stage. Both the voltage envelopes and the course of the photovoltage show good reproducibility for the individual experimental repetitions within the same electrolyte. A direct comparison of the electrical, photoelectrical and optical process data between the experiments in the two different electrolytes additionally shows that there are no major differences in the process behavior. Only the process initiation is delayed by a few seconds in the Zr electrolyte, which should result in generally similar coatings. Thus, in principle, the produced sample material is suitable for direct comparisons with regard to mechanical characterizations.

3.3. Microstructure and Phase Analyses

Figure 3 shows the XRD diagrams of the PEO coatings produced under alternating-current (AC) mode in the Zr electrolyte for different process times. For comparison, the diffraction diagram of the reference coating is also shown.

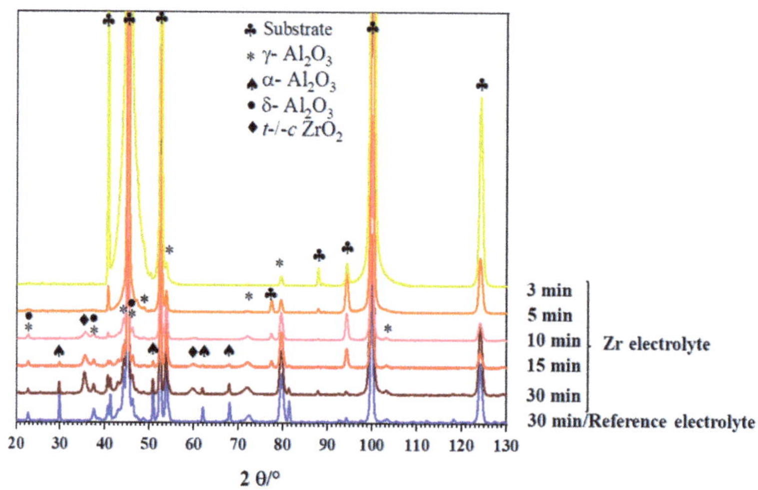

Figure 3. XRD diagrams of coatings prepared from the Zr electrolyte during different process times compared to the reference coating at 30 min process time.

The coating produced in the reference electrolyte is mainly composed of γ-Al_2O_3 (JCPDS card number: 01-074-2206) and α-Al_2O_3 (JCPDS card number: 01-046-1212) as well as δ-Al_2O_3 (JCPDS card number: 00-056-1186). By adding the Zr-based compound to the PEO bath, diffraction peaks of α-Al_2O_3 become weaker, while the peaks related to t/c-ZrO_2 phases (JCPDS card number: 00-050-1089) appear. In order to study the evolution of the phase composition, PEO composite coatings were prepared using different process times (3 min, 5 min, 10 min, and 15 min). As can be seen in Figure 3, the increase in the process time causes the formation of other alumina phases including δ-Al_2O_3 and α-Al_2O_3 in addition to γ-Al_2O_3, similar to the coatings from the reference electrolyte. In addition, t/c-ZrO_2 is detectable in the coatings for processing times longer than 10 min. Figure 4a,b present cross-sectional SEM images using the backscattered electrons detector (BSD) of the Al_2O_3 and Al_2O_3-ZrO_2 coatings, respectively.

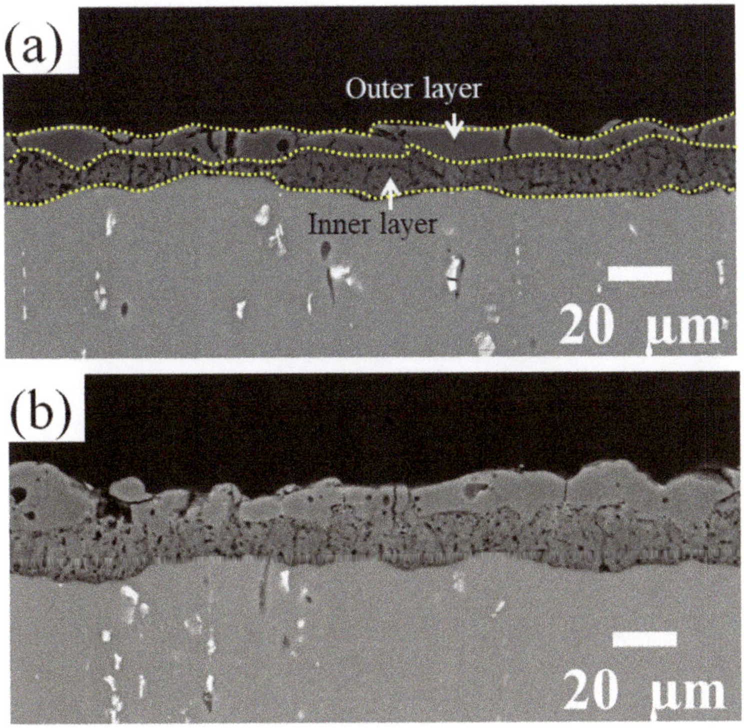

Figure 4. BSD cross-section images of (**a**) the reference coating and (**b**) the coating prepared in the Zr electrolyte. Both PEO coatings exhibit a two-layered structure with a more compact outer layer.

Both coatings have a two-layered structure, which includes a more compact outer layer with vertical cracks, and micro-pores and a finely porous inner layer. However, with the addition of Zr-based salt, the number of vertical cracks in the outer layer is decreased, and they are shallower. The thickness of the layers is presented in Figure 5. The addition of Zr-based salt into the PEO bath increased the thickness of the coating's outer layer from 10 µm to 12 µm and that of the inner layer from 8 µm to 11 µm when using identical parameters in the PEO process.

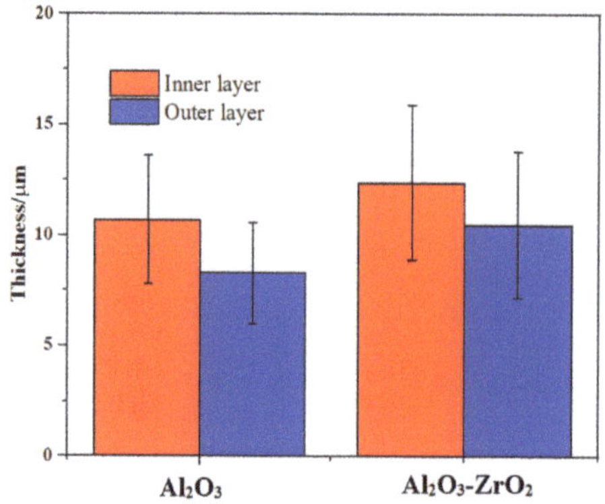

Figure 5. Average value and standard deviation of the thickness of the outer and inner layers of the reference coating and the coating prepared in the Zr electrolyte extracted from the cross-section images in Figure 4. The formed composite coating exhibited a higher thickness for identical processing time, when compared to the reference coating.

EDS analyses were carried out on the cross-sections of the coatings. However, due to the overlapping of Zr L_α and P K_α lines, it is only possible to detect the existence of the two elements with EDS but not to distinguish between them. Therefore, in addition, FESEM was used to identify the Zr distribution using a BSD and the contrast in the micrograph resulting from the difference in the atomic numbers ($Z_P = 15$ and $Z_{Zr} = 40$).

The relatively light-gray regions of the coating in the BSD images indicate an increased amount of Zr relative to the adjacent material. In Figure 6b,d, the detailed examination of the outer layer reveals either zones with a more lamellar structure such as region 1 or zones with a typical dendritic structure such as region 2.

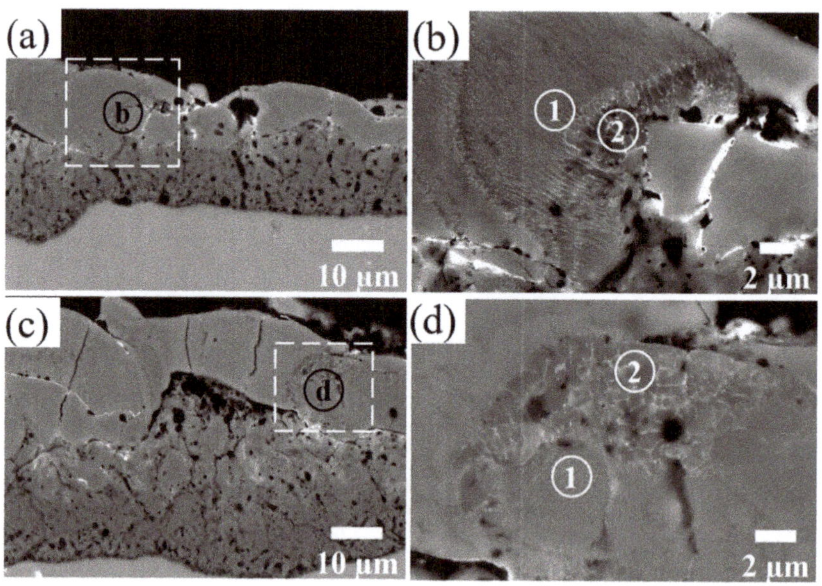

Figure 6. BSD images of the cross-section of the Al_2O_3-ZrO_2 coating with different magnifications. Images (**b**,**d**) show the magnified areas of images (**a**,**c**), respectively. Regions containing Zr either have lamellar (regions 1) or dendritic (regions 2) structures.

3.4. Mechanical Testing

3.4.1. Nano-Indentation Tests

Determined via nano-indentation on the outer layer of the polished cross-sections, the hardness values of the reference and the Al_2O_3-ZrO_2 coatings were (16 ± 4) GPa and (18 ± 2) GPa, respectively. Furthermore, Young's moduli of the reference and Al_2O_3-ZrO_2 coatings were (178 ± 29) GPa and (208 ± 14) GPa, respectively. Therefore, coatings prepared in the Zr electrolyte show higher hardness and higher Young's modulus with a lower standard deviation in comparison with the reference coating.

3.4.2. Nano-Scratch Test

Figure 7 shows representative nano-scratch results and BSD images for the reference and the Al_2O_3-ZrO_2 coatings.

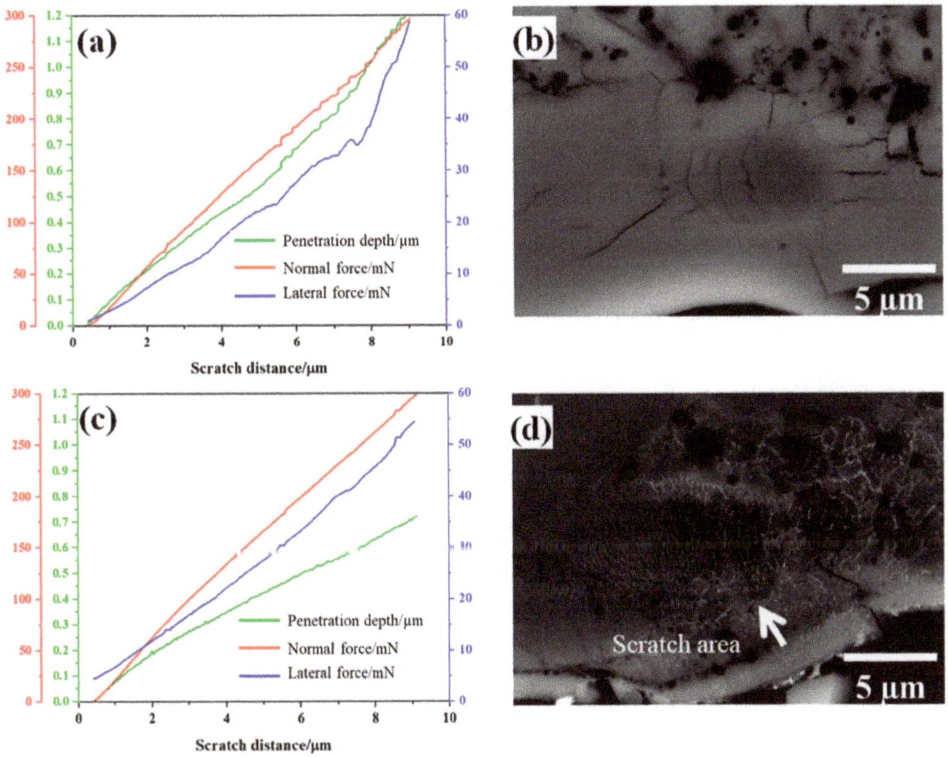

Figure 7. Nano-scratch results and BDS images of one representative scratch on the Al_2O_3 reference coating (**a**,**b**) and the Al_2O_3-ZrO_2 coating (**c**,**d**).

Based on Equation (4), K_c can be calculated through the linear fitting of the F_l (d_p/R) curve. The results are presented in Figure 8. According to the linear behavior of the lateral force vs. (d_p/R), an appropriate fitting range for each scratch was determined. It has to be noted that the data under small and large loads should be discarded in the curve F_l (d_p/R). Under small loads, elastic–plastic deformation occurs, and the cracking plane is not well-developed. When subjected to large loads, the coating failure becomes severe and the cracking behavior becomes complex, thereby rendering the assumption of a semi-circular plane invalid. [24].

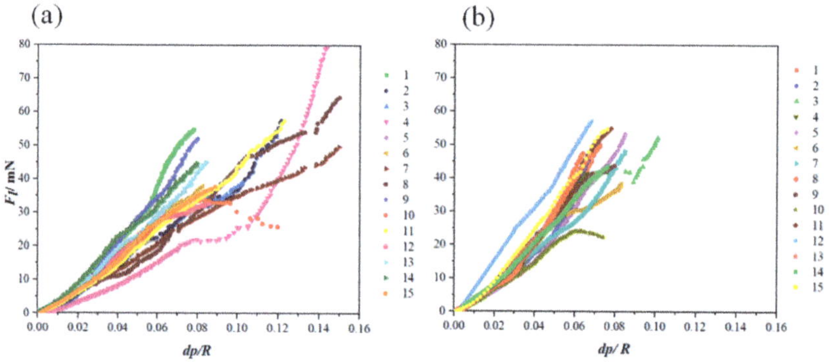

Figure 8. F_l (d_p/R) curves extracted from the nano-scratch results applied on (**a**) the reference coatings and (**b**) the Al_2O_3-ZrO_2 coatings.

Figure 9 shows F_l (d_p/R) curves and the fitting curves of the scratches shown in Figure 7. In addition, Tables 2 and 3 summarize the fitting range, slope of fitting curve for each scratch and the calculated local fracture toughness for the reference coating and the coating prepared from Zr-containing electrolyte. As seen in Figures 7–9, the normal force of 300 mN led to a deeper penetration into the Al_2O_3 coating in comparison with the Al_2O_3-ZrO_2 coating. Therefore, the scratch resulted in severer damage in the Al_2O_3 coating when compared to the Al_2O_3-ZrO_2 coating. This is also clearly noticeable in the BSD images of the scratches. As seen in Figure 7b, sharp and deep cracks were formed and propagated on the cross-section of the Al_2O_3 coating due to the scratch test. However, the scratch in the Al_2O_3-ZrO_2 coating is barely detectable and no cracks are visible (see Figure 7d). Using this approach, the average fracture toughness of the Al_2O_3 reference and the Al_2O_3-ZrO_2 coatings were calculated and are presented in Figure 10. The incorporation of Zr into alumina caused an increase in the fracture toughness to (6.9 ± 1.5) MPa·$m^{1/2}$ when compared to (4.6 ± 1.1) MPa·$m^{1/2}$ for the reference coating.

Table 2. The fitting range of F_l (d_p/R) curves and the related slopes used for the calculation of the fracture toughness of the Al_2O_3 coatings.

Test Number	Fitting Range (d_p/R)	Slope (Linear Fit)	K_c (MPa·$m^{1/2}$)
(1)	0.02–0.06	636	6.2
(2)	0.01–0.05	407	3.9
(3)	0.01–0.06	478	4.6
(4)	0.01–0.07	295	2.8
(5)	0.01–0.06	450	4.4
(6)	0.01–0.07	480	4.6
(7)	0.01–0.06	339	3.3
(8)	0.01–0.06	283	2.7
(9)	0.02–0.07	712	6.9
(10)	0.02–0.06	473	4.6
(11)	0.01–0.06	458	4.4
(12)	0.01–0.06	497	4.8
(13)	0.01–0.04	516	4.9
(14)	0.01–0.04	660	6.4
(15)	0.01–0.05	444	4.3

Table 3. The fitting range of F_l (d_p/R) curves and the related slopes used for the calculation of the fracture toughness of the Al_2O_3-ZrO_2 coatings.

Test Number	Fitting Range (d_p/R)	Slope (Linear Fit)	K_c ($MPa{\cdot}m^{1/2}$)
(1)	0.02–0.07	819	7.9
(2)	0.02–0.07	468	4.5
(3)	0.01–0.07	642	6.2
(4)	0.03–0.07	855	8.3
(5)	0.02–0.08	763	7.3
(6)	0.01–0.05	589	5.7
(7)	0.01–0.05	498	4.8
(8)	0.03–0.06	1049	10
(9)	0.02–0.07	768	7.4
(10)	0.01–0.06	473	4.5
(11)	0.02–0.06	712	6.9
(12)	0.03–0.06	852	8.2
(13)	0.02–0.05	690	6.7
(14)	0.02–0.08	656	6.3
(15)	0.01–0.07	797	7.7

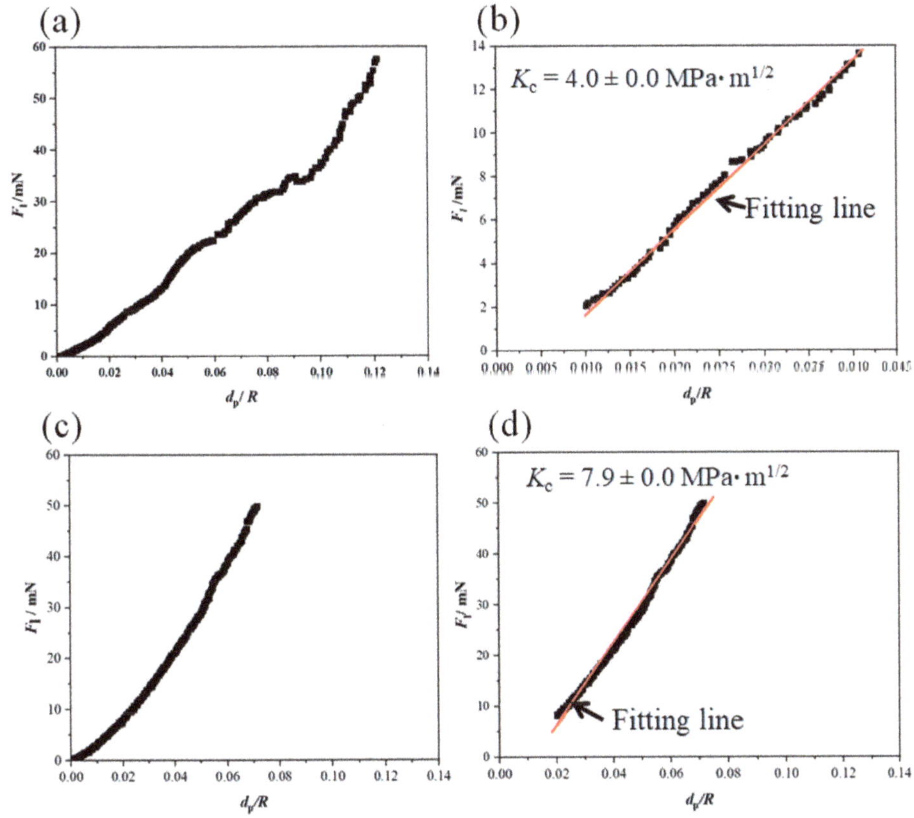

Figure 9. F_l (d_p/R) curve and the related fitting-range curve for the nano-scratch on the Al_2O_3 coating (**a**,**b**) and on the Al_2O_3-ZrO_2 coating (**c**,**d**). These curves were extracted from the nano-scratch results presented in Figure 7.

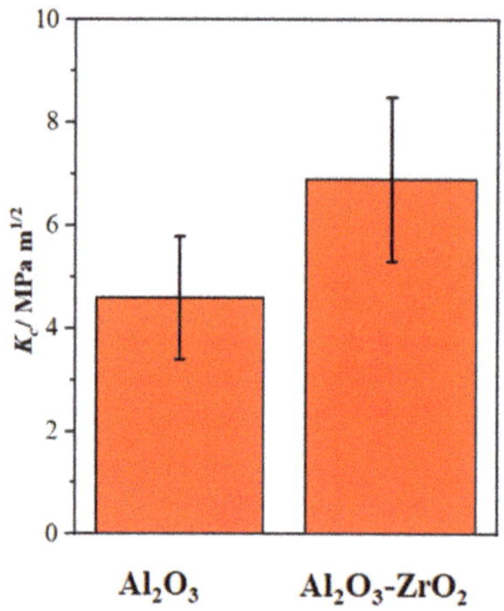

Figure 10. The average fracture toughness with the standard deviation of the Al_2O_3 and the Al_2O_3-ZrO_2 coatings. The Al_2O_3-ZrO_2 coating exhibits significantly higher fracture toughness when compared to the Al_2O_3 reference coating.

4. Discussion

A well-established way for enhancing the fracture toughness of Al_2O_3 is adding ZrO_2 into an alumina matrix. $K_2[Zr(CO_3)_2(OH)_2]$ might be decomposed under thermal plasma condition and produces ZrO_2, which is incorporated into the PEO coating according to [30]:

$$[Zr(CO_3)_2(OH)_2]^{2-} \rightarrow ZrO_2 + 2\,CO_2 \uparrow + 2\,OH^- \quad (5)$$

XRD analyses confirmed that the addition of the Zr-based salt into the PEO bath resulted in the incorporation of a metastable high-temperature modification of zirconia (either t-ZrO_2 or c-ZrO_2) into the alumina matrix, which is composed of various allotropes.

γ-Al_2O_3 and δ-Al_2O_3 are metastable modifications of alumina, and α-Al_2O_3 (corundum) is a thermodynamically stable polymorph of alumina [31]. In the PEO treatment, heat and mass transfers play important roles in the phase formation [32]. Metastable polymorphs, especially γ-Al_2O_3, develop due to high cooling rates of molten material in direct contact with the electrolyte and the substrate [32]. The δ-Al_2O_3 is formed through the structural ordering of γ-Al_2O_3 and its existence indicates that α-Al_2O_3 is developed through a two-step order–disorder structural transformation [33].

Zirconia exhibits three distinct structural polymorphs: m-ZrO_2 from room temperature to 1170 °C, which is the most common one; t-ZrO_2 is stable in the temperature range of 1170–2370 °C and has been widely used in materials science due to its high fracture toughness; and the cubic phase (c-ZrO_2, from 2370 °C to the melting point) known as 'diamond-like' zirconia [34]. As seen in Figure 3, the PEO process resulted in the incorporation of a high-temperature allotrope of zirconia (t- or c-ZrO_2) into the coating. The lack of transformation of t/c-ZrO_2 to m-ZrO_2 in the prepared PEO coating might be owing to the stabilization of t/c-ZrO_2 by the simultaneous in situ formation of Al_2O_3 during the PEO process [6]. In addition, rapid cooling of Al_2O_3-ZrO_2 eutectic mixture could cause the appearance of t- or c-ZrO_2. The incorporation of ZrO_2 into alumina resulted in the formation of dendritic and lamellar structures in the outer layer. In addition, the total thickness of the coating increased by the addition of Zr-based salt for equal process parameters when comparing to the reference electrolyte. As melting and solidification are two main processes involved in PEO, the increase in the growth rate of the coating might be due to the low melting point of the alumina–zirconia mixture in comparison with pure alumina. In addition, the enhancement of the passivability of the Al alloy with Zr-based salt can further increase the growth rate of PEO coating.

As observed, the incorporation of zirconia into the PEO alumina coating enhanced the fracture toughness effectively. In addition, the fracture toughness of the prepared Al_2O_3 and Al_2O_3-ZrO_2 coatings is comparable with the bulk Al_2O_3 (4.40 $MPa \cdot m^{1/2}$) and bulk Al_2O_3-ZrO_2 (6–12 $MPa \cdot m^{1/2}$), respectively [35]. This indicates that the PEO process is an effective way for the fabrication of ZTA coatings, especially since the PEO process results in the stabilization of a high-temperature allotrope of zirconia (*t*- or *c*-ZrO_2) at room temperature without the need for rare-earth dopants such as Y_2O_3. The improvement of the mechanical parameters including hardness and fracture toughness might be related to the effect of zirconia on the microstructural properties of alumina. As seen in SEM images (Figure 6), the incorporation of zirconia developed new morphologies including dendritic and lamellar ones. Adding zirconia to alumina can lead to microstructural refinement, which results in the enhancement of the strength of the ceramic. In addition, zirconia may segregate along alumina grain boundaries and hinder the formation of cracks or promote crack deflection and crack pinning [4]. The other well-established toughening mechanism of alumina caused by zirconia is that *t*-ZrO_2 could be transformed to *m*-ZrO_2 due to stress induced by the interaction of the crack with ZrO_2. This transformation is accompanied by a volume expansion and could hinder the crack propagation [36].

Furthermore, it could be shown that nano-scratch tests can be utilized as an accurate and powerful technique for the evaluation of the fracture toughness of ceramic coatings. Here, fracture toughness is determined directly by the scratch results and there is no need for the measurement of crack lengths in the four corners of the indent, which is a challenge in inherently crack-containing materials such as PEO coatings.

5. Summary and Conclusions

$K_2[Zr(CO_3)_2(OH)_2]$ salt was added to an aluminate-phosphate-based electrolyte to form an Al_2O_3-ZrO_2 coating by a PEO process. The prepared coating had a double-layer microstructure with a finely porous inner layer and a dense outer layer. The addition of the Zr-based salt enhanced the growth rate and improved the integrity of the outer layer. This was attributed to the improvement of passivablity of the substrate by the Zr-based salt. New morphologies (lamella and dendrites) appeared in the coating prepared from the Zr electrolyte. Zirconia was incorporated as a metastable high-temperature modification into the alumina coating. Nano-indentation test indicated that Zr-based salt led to an increase in the hardness of the outer layer from 16 GPa to 18 GPa. In addition, by applying nano-scratch tests on the outer layer, an increase in the local fracture toughness from 4.6 $MPa \cdot m^{1/2}$ to 6.9 $MPa \cdot m^{1/2}$ occurred. Regarding the comparability of the fracture toughness results with the literature, it was shown that the nano-scratch method could be an appropriate replacement for nano- or micro-indentation techniques for the evaluation of fracture toughness in materials with inherent cracks and porosity. Further research must be carried out focusing on a more homogenous distribution of the formed zirconia phases to achieve ZTA coatings produced by PEO with an excellent bonding to the aluminum substrate.

Author Contributions: Conceptualization, M.H., F.S. and L.W.; data curation, M.H.; methodology, M.H., F.S. and L.W.; validation, M.H., F.S. and L.W.; investigation, M.H.; resources, T.L.; writing—original draft preparation, M.H.; writing—review and editing, F.S., L.W. and T.L.; visualization, M.H.; supervision, T.L.; project administration, T.L.; funding acquisition, T.L. All authors have read and agreed to the published version of the manuscript.

Funding: This research was funded by the German Research Foundation (Deutsche Forschungsgemeinschaft/DFG), Grant No. 500196783, project LA-1274/67-1.

Institutional Review Board Statement: Not applicable.

Informed Consent Statement: Not applicable.

Data Availability Statement: The authors confirm that the data supporting the findings of this study are available within the article.

Conflicts of Interest: The authors declare no conflict of interest.

References

1. Huang, T.; Song, P.; Li, C.; Shu, Y.; Sun, B.; Ji, Q.; Arif, M.; Yi, J. Phase transition and interface evolution of Al_2O_3/ZrO_2 particles in plasma-sprayed coatings. *Ceram. Int.* **2020**, *46*, 12275–12281. [CrossRef]
2. Matykina, E.; Arrabal, R.; Skeldon, P.; Thompson, G.E. Incorporation of zirconia nanoparticles into coatings formed on aluminium by AC plasma electrolytic oxidation. *J. Appl. Electrochem.* **2008**, *38*, 1375–1383. [CrossRef]
3. Tarasi, F.; Medraj, M.; Dolatabadi, A.; Oberste-berghaus, J.; Moreau, C. Structural considerations in plasma spraying of the alumina—Zirconia composite. *Surf. Coat. Technol.* **2011**, *205*, 5437–5443. [CrossRef]
4. Pappas, J.M.; Thakur, A.R.; Dong, X. Effects of zirconia doping on additively manufactured alumina ceramics by laser direct deposition. *Mater. Des.* **2020**, *192*, 108711. [CrossRef]
5. Rehman, Z.U.; Shin, S.H.; Kaseem, M.; Uzair, M.; Koo, B.H. Towards a compact coating formed on Al6061 alloy in phosphate based electrolyte via two-step PEO process and K_2ZrF_6 additives. *Surf. Coat. Technol.* **2017**, *328*, 355–360. [CrossRef]
6. Wang, P.; Li, J.; Guo, Y.; Yang, Z.; Wang, J. In situ formation of Al_2O_3–ZrO_2–Y_2O_3 composite ceramic coating by plasma electrolytic oxidation on $ZAlSi_{12}Cu_3Ni_2$ alloy. *Mater. High Temp.* **2018**, *35*, 363–370. [CrossRef]
7. Barati, N.; Yerokhin, A.; Golestanifard, F.; Rastegari, S.; Meletis, E.I. Alumina-zirconia coatings produced by Plasma Electrolytic Oxidation on Al alloy for corrosion resistance improvement. *J. Alloys Compd.* **2017**, *724*, 435–442. [CrossRef]
8. Sieber, M.; Simchen, F.; Morgenstern, R.; Scharf, I.; Lampke, T. Plasma electrolytic oxidation of high-strength aluminium alloys—Substrate effect on wear and corrosion performance. *Metals* **2018**, *8*, 356. [CrossRef]
9. Yerokhin, A.L.; Voevodin, A.A.; Lyubimov, V.V.; Zabinski, J.; Donley, M. Plasma electrolytic fabrication of oxide ceramic surface layers for tribotechnical purposes on aluminium alloys. *Surf. Coat. Technol.* **1998**, *110*, 140–146. [CrossRef]
10. Simchen, F.; Sieber, M.; Lampke, T. Electrolyte influence on ignition of plasma electrolytic oxidation processes on light metals. *Surf. Coat. Technol.* **2017**, *315*, 205–213. [CrossRef]
11. Kaseem, M.; Kwon, J.H.; Ko, Y.G. Modification of a porous oxide layer formed on an Al-Zn-Mg alloy via plasma electrolytic oxidation and post treatment using oxalate ions. *RSC Adv.* **2016**, *6*, 107109–107113. [CrossRef]
12. Student, M.; Pohrelyuk, I.; Padgurskas, J.; Posuvailo, V.; Hvozdets' kyi, V.; Zadorozhna, K.; Chumalo, H.; Veselivska, H. Kovalchuk, I.; Kychma, A. Influence of Plasma Electrolytic Oxidation of Cast Al-Si Alloys on Their Phase Composition and Abrasive Wear Resistance. *Coatings* **2023**, *13*, 637. [CrossRef]
13. Rodriguez, L.; Paris, J.-Y.Y.; Denape, J.; Delbé, K. Micro-Arcs Oxidation Layer Formation on Aluminium and Coatings Tribological Properties—A Review. *Coatings* **2023**, *13*, 373. [CrossRef]
14. Akatsu, T.; Kato, T.; Shinoda, Y.; Wakai, F.; Access, O. Zirconia based ceramic coating on a metal with plasma electrolytic oxidation In Proceedings of the IOP Conference Series: Materials Science and Engineering; IOP Publishing: Bristol, UK, 2011; Volume 18, p 202005.
15. Alloy, A.; Xie, X.; Yang, E.; Zhang, Z.; Wei, W.; Li, D.; Zhao, X.; Yang, R. Effects of K_2TiF_6 and Electrolyte Temperatures on Energy Consumption and Properties of MAO Coatings on 6063. *Materials* **2023**, *16*, 1830.
16. Akatsu, T.; Kato, T.; Shinoda, Y.; Wakai, F. Thermal barrier coating made of porous zirconium oxide on a nickel-based single crystal superalloy formed by plasma electrolytic oxidation. *Surf. Coat. Technol.* **2013**, *223*, 47–51. [CrossRef]
17. Lee, B.; Kwon, J.-S.; Khalid, M.W.; Kim, K.-M.; Kim, J.; Lim, K.M.; Hong, S.H. Boron nitride nanoplatelets as reinforcement material for dental ceramics. *Dent. Mater.* **2020**, *36*, 744–754. [CrossRef]
18. Pinto, P.A.; Colas, G.; Filleter, T.; De Souza, G.M. Surface and mechanical characterization of dental yttria-stabilized tetragonal zirconia polycrystals (3Y-TZP) after different aging processes. *Microsc. Microanal.* **2016**, *22*, 1179–1188. [CrossRef]
19. Chen, J. Indentation-based methods to assess fracture toughness for thin coatings. *J. Phys. D Appl. Phys.* **2012**, *45*, 203001. [CrossRef]
20. Karimzadeh, A.; Ayatollahi, M.R. Mechanical properties of biomaterials determined by nano-indentation and nano-scratch tests. In *Nanomechanical Analysis of High Performance Materials*; Springer: Berlin, Germany, 2014; pp. 189–207.
21. Sebastiani, M.; Johanns, K.E.; Herbert, E.G.; Pharr, G.M. Measurement of fracture toughness by nanoindentation methods: Recent advances and future challenges. *Curr. Opin. Solid State Mater. Sci.* **2015**, *19*, 324–333. [CrossRef]
22. Akono, A.; Randall, N.X. Experimental determination of the fracture toughness via microscratch tests: Application to polymers, ceramics, and metals. *J. Mater. Res.* **2011**, *27*, 485–493. [CrossRef]
23. Hashemzadeh, M.; Raeissi, K.; Ashrafizadeh, F.; Hakimizad, A.; Santamaria, M. The incorporation mechanism of colloidal TiO_2 nanoparticles and their effect on properties of the coatings grown on 7075 Al alloy from silicate-based solution using plasma electrolytic solution. *Trans. Nonferrous Met. Soc. China* **2021**, *31*, 3659–3676.
24. Liu, M.; Xu, Z.; Fu, R. Micromechanical and microstructure characterization of BaO-Sm_2O_3-$5TiO_2$ ceramic with addition of Al_2O_3. *Ceram. Int.* **2022**, *48*, 992–1005. [CrossRef]
25. Zhang, D.; Sun, Y.; Gao, C.; Liu, M. Measurement of fracture toughness of copper via constant-load microscratch with a spherical indenter. *Wear* **2020**, *444–445*, 203158. [CrossRef]
26. Simchen, F.; Masoud-Nia, N.; Mehner, T.; Lampke, T. Formation of corundum-rich alumina coatings on low-carbon steel by plasma electrolytic oxidation. In Proceedings of the IOP Conference Series: Materials Science and Engineering; IOP Publishing: Bristol, UK, 2021; Volume 1147, p. 12007.

Simchen, F.; Sieber, M.; Mehner, T.; Lampke, T. Characterisation Method of the Passivation Mechanisms during the pre-discharge Stage of Plasma Electrolytic Oxidation indicating the Mode of Action of Fluorides in PEO of Magnesium. *Coatings* **2020**, *10*, 965. [CrossRef]

Oliver, W.C.; Pharr, G.M. Measurement of hardness and elastic modulus by instrumented indentation: Advances in understanding and refinements to methodology. *J. Mater. Res.* **2004**, *19*, 3–20. [CrossRef]

Huang, L.Y.; Zhao, J.W.; Xu, K.W.; Lu, J. A new method for evaluating the scratch resistance of diamond-like carbon films by the nano-scratch technique. *Diam. Relat. Mater.* **2002**, *11*, 1454–1459. [CrossRef]

Ryu, T.; Joon Choi, Y.; Hwang, S.; Sohn, H.Y.; Kimw, I. Synthesis of yttria-stabilized Zirconia Nanopowders by a Thermal Plasma Process. *J. Am. Ceram. Soc.* **2010**, *93*, 3130–3135. [CrossRef]

Ruppi, S. Deposition, microstructure and properties of texture-controlled CVD α-Al_2O_3 coatings. *Int. J. Refract. Met. Hard Mater.* **2005**, *23*, 306–316. [CrossRef]

Belozerov, V.; Sobol, O.; Mahatilova, A.; Subbotina, V.; Taha, A.T.; Ubeidulla, F.; Safwan, M.; Al-qawabah, S.M.; Belozerov, V.; Sobol, O.; et al. Effect of electrolysis regimes on the structure and properties of coatings on aluminum alloys formed by anode-cathode micro arc oxidation. *Mater. Sci.* **2018**, *1*, 43–47. [CrossRef]

Levin, I.; Brandon, D. Metastable alumina polymorphs: Crystal structures and transition sequences. *J. Am. Ceram. Soc.* **1998**, *81*, 1995–2012. [CrossRef]

Pimentel, H.R.X.; Aguiar, D.L.M.; San Gil, R.A.S.; Souza, E.F.; Ferreira, A.R.; Leitão, A.A.; Alencastro, R.B.; Menezes, S.M.C.; Chiaro, S.S.X. 17O MAS NMR and first principles calculations of ZrO_2 polymorphs. *Chem. Phys. Lett.* **2013**, *555*, 96–100. [CrossRef]

Abbas, S.; Maleksaeedi, S.; Kolos, E.; Ruys, A.J. Processing and properties of zirconia-toughened alumina prepared by gelcasting. *Materials* **2015**, *8*, 4344–4362. [CrossRef] [PubMed]

De Aza, A.H.; Chevalier, J.; Fantozzi, G.; Schehl, M.; Torrecillas, R. Crack growth resistance of zirconia toughened alumina ceramics for joint prostheses. *Key Eng. Mater.* **2002**, *23*, 937–945. [CrossRef]

isclaimer/Publisher's Note: The statements, opinions and data contained in all publications are solely those of the individual ıthor(s) and contributor(s) and not of MDPI and/or the editor(s). MDPI and/or the editor(s) disclaim responsibility for any injury to eople or property resulting from any ideas, methods, instructions or products referred to in the content.

coatings

MDPI

Article

The Effect of W, Cr, Mo Content on the Microstructure and Mechanical Properties of the Weld Interface of TiC Cermet and Low-Carbon Steel

Wei Wei [1,2], Zhiquan Huang [1], Haiyan Zhang [1] and Shaokang Guan [2,*]

[1] Zhengzhou Research Institute of Mechanical Engineering, Zhengzhou 450000, China; weiw@zrime.com.cn (W.W.); tehanshiyanshi@hotmail.com (Z.H.); trunks.ww@163.com (H.Z.)
[2] School of Material Science & Enginreering, Zhengzhou University, Zhengzhou 450000, China
* Correspondence: skguan@zzu.edu.cn

Abstract: In this study, the influence of W, Cr, and Mo on the microstructure and mechanical properties of the arc-welded interface of TiC cermet and low-carbon steel was investigated. MIG arc welding was employed to deposit muti-alloyed low-carbon steel flux-cored wire onto the surface of the TiC cermet to create the arc-welded interface. Analysis of the microstructure, phase composition, and shear fracture of the interface were conducted by OM (optical microscopy), SEM (scanning electron microscope), EMPA (Electron Probe X-ray Micro-Analyzer), and XRD (X-ray diffraction) methods. The results indicate that the order of influence on the performance of the welded interface is perceived as Cr > W > Mo. The preferred ratio of element content is W at 1.0 wt.%, Cr at 0.5 wt.%, and Mo at 2.0 wt.%. During the arc-welding process, W and Mo formed a rim structure of TiC particles to inhibit the dissolution of TiC particles, while Cr formed dispersed carbides in the bonding phase. The synergistic impact of these components resulted in the simultaneous enhancement of both the TiC particles and the bonding phase. This led to a significant increase in the shear strength of the TiC cermet welded interface to 787 MPa, marking an 83% improvement compared to the welded interface without reinforcement, which exhibited a shear strength of 430 MPa.

Keywords: TiC cermet; arc welding; multi-alloy strengthening; interface; shear strength

1. Introduction

TiC cermet is primarily composed of TiC or Ti(C,N) as the main components, with elements such as Ni, Mo, and Fe used as the main elements to create the bonding phase. Compared to WC particles, TiC particles have the characteristics of low density (4.93 g/cm^3), high hardness (3200 HV) [1], and stable properties. The literature shows that the wear resistance of the matrix reinforced with TiC is increased by 1–2 orders compared to non-reinforcement [2]. TiC cermet inherits the advantages of both cemented carbide and steel, combining the high hardness and wear resistance of cemented carbide with the machinability, heat treatability, and forgeability of steel, demonstrating excellent overall performance [3–6]. However, the manufacturing cost of TiC cermet is relatively high, the manufacturing conditions are strict, and there are difficulties in producing large-sized parts, leading to its limited application range due to its expensive price [7,8]. Therefore, the majority of successful applications depend on the bond between the TiC cermet and the metal matrix.

The weldability of TiC cermet was investigated based on previous research achievements [9]. A Gleeble simulation was employed to occur the heat affect zone of TiC cermet, and the microstructure and mechanical properties results showed that TiC cermet has a stable chemical and physical properties during the welding heat-input process; hence, the TiC cermet has certain weldability. The research conducted focused on identifying the most suitable weld metal for welding TiC cermet. Various alloyed welding materials were used to study the microstructure and mechanical properties of the welded interface in TiC cermet. The findings revealed that low-carbon alloy steel with ferrite is the most suitable welding material for arc welding TiC cermet. This suggests that using this specific material can result in improved weld quality and overall performance of the welded interface [10]. Therefore, the reinforcement of the bond between the TiC cermet and the metal matrix is being valued. Lee [11] studied a TiC metal ring containing Mo_2C and WC, and this specific microstructure inhibited the diffusion of carbides in the core (TiC) and edge (Ti, Mo, or W). This phenomenon combines the effects of solid-state diffusion processes (inner edge) and dissolution–precipitation mechanisms (outer edge). The addition of W to TiC gives it a smaller lattice constant compared to other elements, resulting in slightly larger lattice vibrations [12]. Lin [13] investigated the effects of chromium on TiC cermet; the results show that with an increase in the chromium content, the porosity of sintered samples increases and the TRS of the sample gradually decreases, with the hardness firstly increasing and then decreasing a little.

The literature shows that TiC cermet has certain weldability and that low-carbon alloy steel is most suitable for arc welding, while W, Cr, and Mo alloy elements can inhibit the decomposition of TiC particles and improve the strength of sintering test samples in the liquid sintering process of TiC steel cemented carbide. In this article, the MIG welding method was employed to obtain weld interfaces between low-carbon steel and TiC cermet. W, Cr, and Mo were added to the low-carbon steel weld material to enhance the interface. The analysis of the microstructure and mechanical properties of the joints was conducted to identify the preferred Cr, W, and Mo content.

2. Materials and Methods

Materials

An orthogonal experimental table of three factors and four levels was adopted to investigate the effects of W, Cr, and Mo on the TiC cermet welding interface. The low-carbon steel welding wire and the TiC cermet used in this study were produced by Zhengzhou Research Institute of Mechanical Engineering Co., Ltd., Zhengzhou, China, and the chemical and mechanical properties tested following «GB/T 223» [14] «GB/T 40388» [15] and «ASTM E1269-11» [16] standards, respectively. The results are shown in Table 1.

Table 1. Chemical (wt.%) and mechanical properties of weld metal and TiC cermet.

	C	Ni	Mn	TiC	Mo	Fe	Shear Strength (MPa)
ZDZC60	0.53	1.78	6.21	39.50	1.31	etc.	469
LCS	0.051	-	0.78	-		etc.	430

The alloys' content levels were set at 0.5 wt.%, 1.0 wt.%, 1.5 wt.%, and 2.0 wt.%. Since the original low-carbon alloy steel contains 0.5 wt.% of Mo, the Mo content was set at 1.0 wt.%, 1.5 wt.%, 2.0 wt.%, and 2.5 wt.%. Table 2 shows the specific experimental arrangement of the orthogonal experimental table with three factors and four levels. The composition of the welded metal and TiC cermet is shown in Table 2. The welding parameters were set as follows: welding current of 270 A, voltage of 27 V, and welding speed of 0.8 cm/s. The microhardness gradients of the joints were obtained by measuring at intervals of 0.1 mm perpendicular to the interface. The morphology of the samples was analyzed by Phoneme XL1 scanning electron microscope (SEM) and EPMA. The phase patent was analyzed by D8 type X-ray diffraction (XRD).

Table 2. Element design.

	W	Cr	Mo
1#	0.5	0.5	1.0
2#	0.5	1.0	1.5
3#	0.5	1.5	2.0
4#	0.5	2.0	2.5
5#	1.0	0.5	1.5
6#	1.0	1.0	1.0
7#	1.0	1.5	2.5
8#	1.0	2.0	2.0
9#	1.5	0.5	2.0
10#	1.5	1.0	2.5
11#	1.5	1.5	1.0
12#	1.5	2.0	1.5
13#	2.0	0.5	2.5
14#	2.0	1.0	2.0
15#	2.0	1.5	1.5
16#	2.0	2.0	1.0

3. Results

The orthogonal experimental analysis was conducted on the samples of low-carbon alloy steel welded with TiC steel cemented carbide, and the analysis of the shear strength data of the low-carbon alloy steel welded with TiC cermet arc welding in the orthogonal experiment is shown in Table 3.

Table 3. Data analysis of shear strength (MPa) of TiC cermet arc-welding interface.

	W	Cr	Mo	Shear Strength (MPa)
1#	0.5	0.5	1.0	746
2#	0.5	1.0	1.5	712
3#	0.5	1.5	2.0	698
4#	0.5	2.0	2.5	765
5#	1.0	0.5	1.5	752
6#	1.0	1.0	1.0	711
7#	1.0	1.5	2.5	773
8#	1.0	2.0	2.0	767
9#	1.5	0.5	2.0	782
10#	1.5	1.0	2.5	646
11#	1.5	1.5	1.0	761
12#	1.5	2.0	1.5	747
13#	2.0	0.5	2.5	751
14#	2.0	1.0	2.0	727
15#	2.0	1.5	1.5	755
16#	2.0	2.0	1.0	714
mean value 1	730.25	757.75	733.00	
mean value 2	750.75	699.00	741.50	
mean value 3	734.00	746.75	743.50	
mean value 4	736.75	748.25	733.75	
MAX	750.75	757.75	743.50	
MIN	730.25	699.00	733.00	
range	20.50	58.75	10.50	

A comparison of the effects of each alloying element on the shear strength of the welded TiC cermet interface reveals that the element Cr has the most significant impact, with a range of 58.75. The maximum mean value is 757.75, corresponding to the level of 0.5 wt.%. The element W has the second most significant impact, with a range of 20.50. Its maximum value is 750.75, corresponding to the level of 1.0 wt.%. The impact of the element Mo is the least significant, with a range of 10.50. Its maximum value is 743.50, corresponding to the level of 2.0 wt.%. In terms of the shear strength performance, the optimal alloy composition should involve W content of 1.0 wt.%, Cr content of 0.5 wt.%, and Mo content of 2.0 wt.%.

For verification purposes, a welding wire with the optimal alloy composition was prepared. The composition of the weld metal is shown in Table 4.

Table 4. Composition of weld metal (wt.%).

	C	W	Mn	Cr	Mo	Fe
LCS	0.051	0.95	0.78	0.55	1.94	etc.

The low-carbon alloy steel welding wire, with W, Cr, and Mo alloy element strengthening ratios of 1.0 wt.%, 0.5 wt.%, and 2.0 wt.%, respectively, was used to arc weld the interface of TiC cermet, and the microstructure was observed using SEM. Figure 1 illustrates the microstructure results of the weld interface between the TiC cermet and the welding materials. Remarkably, the interface exhibited successful metallurgical bonding with TiC cermet.

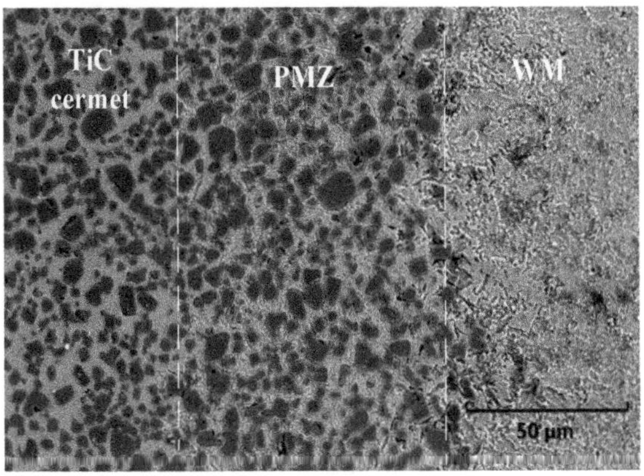

Figure 1. SEM microstructure at the interface.

The arc-welded interface maintains the heat-affected zone (HAZ)–partially melted zone (PMZ)–unmelted zone (UZ)–weld metal (WM) microstructure. The arc-welding interface is observed by SEM, as shown in Figure 1. The TiC cermet and the metal exhibited good metallurgical bonding at the interface, and no microcracks or defects were found at the weld interface compared to the interface welded with γ-Fe or γ-Ni weld metal in previous research [10], indicating good metallurgical bonding between the weld metal and the TiC cermet. A comparison with the analysis of non-alloy-strengthened welding interfaces in [10] reveals that multi-alloy strengthening inhibited the decomposition of TiC particles, resulting in only a small amount of in-situ carbides being formed in the UZ. At the same time, multi-alloy strengthening promotes inter-diffusion between the overlay metal and the TiC cermet, enlarging the PMZ, with a PMZ width of 74 μm, more wild than the width of 20 μm with none strength weld interface [10].

Figure 2 compared the EPMA results of the PMZ between the multi-alloy-strengthened and non-strengthened samples. The microstructure of the PMZ is shown in Figure 2a,b, with the black TiC particle diffusion distributed in the light gray bonding phase. An obvious dark black inner core surrounded by a lighter black rim structure was observed in the TiC particles. While a clear rim distribution of W on the outer layer of the TiC particles was detected in the multi-alloy-strengthened PMZ, the TiC core was similarly surrounded by a W-rich solid-solution carbide phase of Ti-W-C. In contrast, little W was detected in the non-strengthened PMZ, as shown in Figure 2c,e. During the melting process, the TiC particles dissolved into a solid–liquid phase, in which the mass transport of the carbides are driven by W and Mo, as both W and Mo are strong carbide-forming elements. The formation of the 'core-rim' structure is able to increase the wetting of TiC by the bonding phase in order to improve the bonding strength between TiC particles and the bonding phase. There was no obvious clustering of Cr in either sample; as shown in Figure 2d,f, the Cr element is dispersed throughout the bonding phase, and no Cr-rich solid-solution carbide rim structure was observed. It can be observed in Figure 2g,i that in both samples there is a noticeable accumulation of Mo on the outer layer of the TiC particles. During the cooling process, a 'core-rim' structure of Mo-rich solid-solution carbide formed the Ti-Mo-C phase as the outer layer rim surrounding the TiC core. This structure is similar to that of TiC particles in sintered TiC cermet [17].

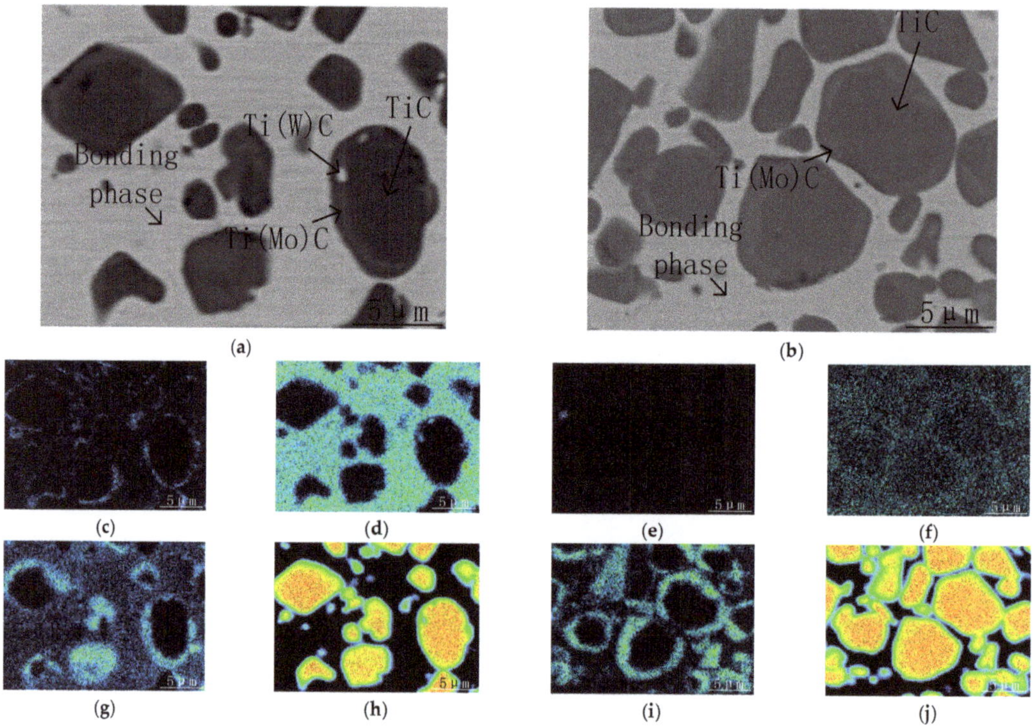

Figure 2. Element distribution in PMZ. (**a**) Multi-alloy strengthened microstructure; (**b**) non-strengthened microstructure; (**c**) W distribution in (**a**); (**d**) Cr distribution in (**a**); (**e**) W distribution in (**b**); (**f**) Cr distribution in (**b**); (**g**) Mo distribution in (**a**); (**h**) Ti distribution in (**a**); (**i**) Mo distribution in (**b**); (**j**) Ti distribution in (**b**).

Arc welding is a rapid heating and cooling metallurgical reaction process. During the welding process, TiC particles decompose to release C and Ti. In this experiment, due to insufficient time for the C to diffuse into the bonding phase, a metallurgical reaction occurred on the surface of the TiC particles, leading to the formation of a 'core-rim' structure, with the TiC core surrounded by a Ti-(Mo,W)-C phase. While there was no Cr rim structure in Figure 2d, as Mo and W are stronger carbide-forming elements than Cr, the characterization reveals the formation of a compound primarily composed of Mo, W, C, and Ti on the surface of TiC particles, which effectively inhibits the melting decomposition of internal TiC particles [18]. Consequently, after the formation of the 'core-rim' structure primarily composed of Mo, W, C, and Ti, the short amount of C in the bonding phase led to the partial formation of Cr carbide particles and solid-solution strengthening dispersed throughout the bonding phase. XRD analysis of the interface results is shown in Figure 3.

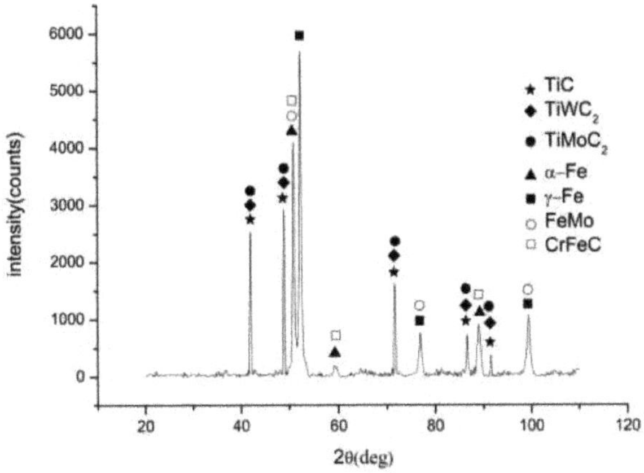

Figure 3. XRD test of the welded interface.

Based on the results shown in Figure 3, the optimal-ratio arc-welded TiC cermet interface contains peaks of TiC, γ-Fe, and α-Fe phases, which is consistent with the phase analysis results of the interface between low-carbon steel and TiC cermet [10]. With the addition of Mo and W, there are peaks of phases containing Ti(Mo,W)C in the interface XRD scan results. Combined with the analysis in Figure 2c,g, a composite carbide of 'core-rim' structure is formed by the Ti(Mo,W)C phase surrounding the TiC particles in the PMZ. The presence of peaks corresponding to Cr-Fe-C, along with the absence of peaks of phases containing Ti-Cr-C, reveals that Cr did not play a role in the formation of the rim structure. Instead, Cr contributed to the formation of carbides with a diffusion distribution in the bonding phase alone with solid-dissolution strengthening.

The shear strength of the interface of the multi-alloy-strengthened low-carbon alloy steel and TiC cermet was tested, and the result was compared with the interface of the TiC cermet arc welded with low-carbon alloy steel welding wire, the non-strengthened arc-welded interface, sintered TiC cermet, and TiC cermet after 6.2 kJ/cm heat input simulation, as shown in Table 5.

Table 5. Verification of the shear strength of TiC cermet interface [9,10].

	Shear Strength (MPa)
Multi-alloy-strengthened	787
Non-strengthened	430
Sintered TiC cermet	469
TiC cermet with 6.2 kJ/cm heat input simulation	584

According to Table 5, the shear strength of the interface significantly increased after being multi-alloy-strengthened compared to other samples. The sample with the optimal ratio had an interface shear strength of 787 MPa, showing a slight improvement compared to the highest value of 782 MPa in the orthogonal experiments (Table 3). Compared to the shear strength of 430 MPa of the non-strengthened interface, an outstanding 83% improvement was occurred through multi-alloy strengthening. When comparing the shear strength of sintered/heat-input-simulated TiC cermet and multi-alloy-strengthened TiC cermet welding interfaces, it was found that the weak point of the weld bond lies on the TiC cermet side. However, it was also observed that the interface between the two materials exhibits a higher shear strength than the TiC cermet itself. This indicates that the multi-alloy-strengthened TiC cermet may offer improved bonding properties and overall performance. The ratios of the elements with the highest shear strength values in the orthogonal experiments were 1.5 wt.%, 0.5 wt.%, and 2.0 wt.%, respectively, which is essentially consistent with the factor content of the optimal ratio.

Figure 4 shows the SEM image of the shear fracture section of the arc-welded interface after strengthening with W, Cr, and Mo alloy elements. It can be observed that there is no significant change in the morphology of the TiC particle phase at the fracture surface of the sample; the shear fracture section is of quasi-cleavage rupture type. The TiC phase contains many differently oriented cleavage steps that exhibit brittle transgranular fractures, while the bonding phase contains a straight tearing ridge. This is attributed to the metal–ceramic characteristics of the TiC particles [19], indicating a strong metallurgical bond between the TiC particles and the bonding phase at the arc-welded interface. As a result, the plastic deformation characteristics become more obvious as the dimples and tearing ridge on the shear fracture section of the sample increased, including shallow small dimples and a few different-sized equiaxed dimples. These deformations signify an improvement in the shear strength of the arc-welded interface of TiC cermet.

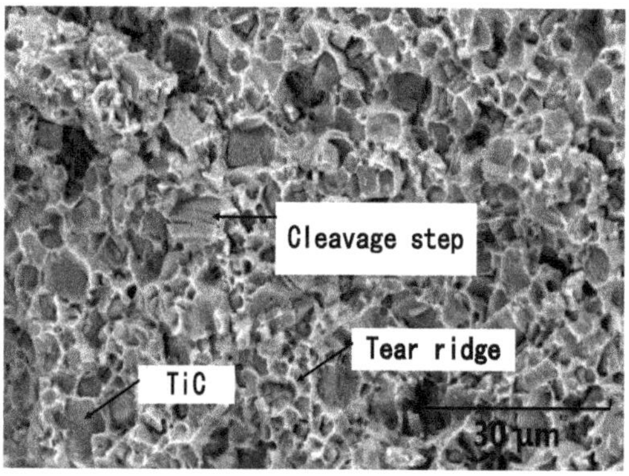

Figure 4. Optimal composition of arc-welded TiC cermet interface fracture.

The fracture section of the samples shows cleavage steps with different orientations. In the shear fracture section, with the diffusion distribution in the bonding phase of Cr carbides and Cr solid-dissolution strengthening in the bonding phase, the toughening dimples of the bonding phase and the tearing ridge of the bonding phase indicate that the TiC cermet bonding phase undergoes plastic deformation before fracturing. The brittle cleavage steps transgranular fractures indicate that during the welding process, TiC particles melted and decomposed, leading to an increase in lattice dislocation, forming carbon-depleted phases, which increased the brittleness of the TiC particles. Hence, secondary cracks were exhibited in the TiC particles. These brittle TiC particles acted as crack initiators, leading to a decrease in the shear strength of the samples. The fracture exhibited quasi-cleavage fractures overall.

Figure 5 illustrates a simplified model depicting the evolution of the microstructure at the interface of low-carbon steel welding material arc welded with TiC cermet following reinforcement with W, Cr, and Mo.

During the welding process, the Ti and C elements in the PMZ formed the shell with the W and Mo elements during the cooling process to inhibit the precipitation of C. Parts of the Cr and C form fine carbides that are distributed in the bonding phase, and part of the Cr contributes to the solid-solution strengthening of the bonding phase.

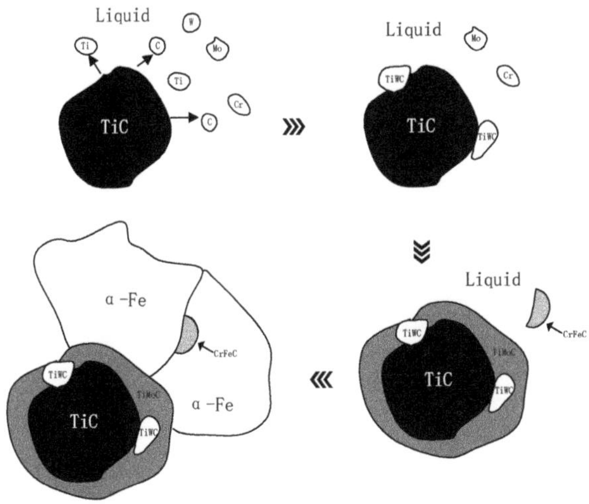

Figure 5. Microstructure evolution of TiC cermet arc welding.

4. Discussion

According to the 'core-rim' theory [20], during the high-temperature stage of welding, TiC particles are distributed in a bonding phase consisting of a mixture of ferrite from the weld metal and austenite from the bonding phase of the TiC cermet, while some TiC particles dissolve into the mixed bonding phase. As W and Mo are stronger carbide-forming elements than Cr, during the cooling process, the W initially nucleates with the C on the surface of the TiC particles, then Mo starts to form the carbides surrounding the TiC particles. With most of the Mo and W participating in reactions on the surface of the TiC particles, a Ti(W,Mo)C outer ring structure forms around the surface of the TiC particles [21], which inhibits the dissolution of the TiC particles, while Cr remains in the mixed bonding phase. The difference from the research conclusions lies in the fact that Jin [22] used a liquid-phase sintering method with a long high-temperature duration of up to 240 min, allowing sufficient time for W, Cr, and Mo elements to form a complex three-layer 'core-rim' structure containing TiC-Ti(W,Mo)C-Ti(Cr,Ni)C. However, arc welding is a process of rapid heating and rapid heat dissipation. Within a brief liquid phase, elements cannot fully react on the surface of TiC particles. Therefore, only W and Mo alloys with stronger carbide-forming elements can nucleate with C on the surface of TiC particles, forming the 'core-rim' structure. Simultaneously, a portion of the Cr reacts with a small quantity of C in the bonding phase, leading to the precipitation of chromium carbides at the ferrite grain boundaries. The presence of dispersed chromium carbides in the matrix enhances the overall strength of the samples. However, as the Cr content increases, the presence of carbide inclusions also escalates. This leads to more C being engaged in the reaction with chromium carbides, with the C originating from the dissolution of TiC particles. Consequently, this process ultimately causes brittleness in the TiC particles. The remaining Cr elements undergo solid dissolution to strengthen the bonding phase. Therefore, the combined effects of Cr strengthening the bonding phase and W and Mo inhibiting the TiC particles collectively influence the mechanical properties at the arc-welded interface of TiC cermet.

The following conclusions can be stated:

(1) An $L^3{}_4$ orthogonal experiment was carried out to investigate the effects of W, Cr, and Mo on the microstructure and properties of the arc-welded interface of TiC cermet. The order of importance within the experimental range on the interface properties was perceived as Cr > W > Mo.

(2) During the welding process, W and Mo alloy elements formed a 'core-rim' structure around the TiC particles, inhibiting the dissolution, while Cr formed dispersed carbides within the bonding phase. The combined effects of these elements enhanced the shear strength of the arc-welded interface of TiC cermet. The optimal alloy proportions for welding TiC cermet were determined to be W 1.0 wt.%, Cr 0.5 wt.%, and Mo 2.0 wt.%.

(3) The shear strength of the welded interface increased by 83% compared to non-strengthened conditions, significantly enhancing the bonding strength of the arc-welded interface of TiC cermet.

Author Contributions: Conceptualization, W.W. and S.G.; methodology, W.W.; validation, W.W., S.G. and Z.H.; investigation, W.W.; writing—review and editing, W.W.; project administration, W.W. and H.Z. All authors have read and agreed to the published version of the manuscript.

Funding: This research received no external funding.

Institutional Review Board Statement: Not applicable.

Informed Consent Statement: Not applicable.

Data Availability Statement: Data are contained within the article.

Acknowledgments: The authors are grateful to Jianjun Wei for giving them encouragement to pursue this work.

Conflicts of Interest: The authors declare no conflicts of interest.

References

1. Kalita, V.I.; Radyuk, A.A.; Komlev, D.I.; Mikhailova, A.B.; Alpatov, A.V.; Chueva, T.R.; Gamurar, N.V.; Titov, D.D. TiC–Cr3C2 WC–TiB2–SiC-Based Cermets. *Inorg. Mater. Appl. Res.* **2020**, *11*, 1199–1213. [CrossRef]
2. SAlgodi, J.; Murray, J.W.; Brown, P.D.; Clare, A.T. Wear performance of TiC/Fe cermet electrical discharge coatings. *Wear* **201** *402–403*, 109–123. [CrossRef]
3. Sribalaji, M.; Mukherjee, B.; Bakshi, S.R.; Arunkumar, P.; Babu, K.S.; Keshri, A.K. In-situ formed graphene nanoribbon induce toughening and thermal shock resistance of spark plasma sintered carbon nanotube reinforced titanium carbide composit *Compos. B Eng.* **2017**, *123*, 227–240. [CrossRef]
4. Xue, J.X.; Liu, J.X.; Zhang, G.J.; Zhang, H.B.; Liu, T.; Zhou, X.S.; Peng, S.M. Improvement in mechanical/physical properties TiC625 ceramics sintered at 1500 °C for inert matrix fuels. *Scr. Mater.* **2016**, *114*, 5–8. [CrossRef]
5. Namini, A.S.; Ahmadi, Z.; Babapoor, A.; Shokouhimehr, M.; Asl, M.S. Microstructure and thermomechanical characteristics spark plasma sintered TiC ceramics doped with nano-sized WC. *Ceram. Int.* **2019**, *45*, 2153–2160. [CrossRef]
6. Babapoor, A.; Asl, M.S.; Ahmadi, Z.; Namini, A.S. Effects of spark plasma sintering temperature on densification, hardness an thermal conductivity of titanium carbide. *Ceram. Int.* **2018**, *44*, 14541–14546. [CrossRef]
7. Maurya, H.S.; Kollo, L.; Tarraste, M.; Juhani, K.; Sergejev, F.; Prashanth, K.G. Selective Laser Melting of TiC-Based Cermet: HI Studies. *Trans. Indian Inst. Met.* **2023**, *76*, 565–570. [CrossRef]
8. Maurya, H.S.; Kollo, L.; Juhani, K.; Sergejev, F.; Prashanth, K.G. Effect of preheating and cooling of the powder bed by laser puls shaping on the microstructure of the TiC based cermets. *Ceram. Int.* **2022**, *48*, 20612–20618. [CrossRef]
9. Wei, W.; Huang, Z.; Zhang, H.; Guan, S. Effect of the Welding Thermal Cycle on the Microstructure and Mechanical Properties TiC Cermet HAZ Using the Gleeble Simulator. *Coatings* **2023**, *13*, 476. [CrossRef]
10. Wei, W.; Huang, Z.; Zhang, H.; Guan, S. The Influence of Different Alloyed Welding Materials on the Microstructure an Mechanical Properties of Welded Interface in TiC Cermet MIG Welding Bonds. *Mater. Lett.* **2024**, *361*, 135641. [CrossRef]
11. Lee, J.H.; Park, H.K. Microstructural evolution and mechanical properties of TiC-Mo_2C-WC-Ni multi-component powder by hig energy ball milling. *J. Ceram. Process. Res.* **2021**, *22*, 590–596.
12. Chen, R.; Chen, L.; Wang, L.; Liang, C. Alloying effects of Zr, Nb, Ta, and W on thermodynamic and mechanical properties of Ti based on first-principles calculation. *Nucl. Mater. Energy* **2024**, *38*, 101604. [CrossRef]
13. Lin, T.; Guo, Y.; Wang, Z.; Shao, H.; Lu, H.; Li, F.; He, X. Effects of chromium and carbon content on microstructure and propertie of TiC-steel composites. *Int. J. Refract. Met. Hard Mater.* **2018**, *72*, 228–235. [CrossRef]
14. *GB/T 223.54.2022*; Iron, Steel and Alloy—Determination of Nickel Content—Flame Atomic Absorption Spectrometric Metho Standards Press of China: Beijing, China, 2022.
15. *GB/T 40388-2021*; Test Method for Shear Strength of C/C Composites. Standards Press of China: Beijing, China, 2022.
16. *ASTM E1269-11*; Standard Test Method for Determining Specific Heat Capacity by Differential Scanning Calorimetry. ASTM West Conshohocken, PA, USA, 2018.

17. Jõeleht, M.; Pirso, J.; Juhani, K.; Viljus, M.; Traksmaa, R. The influence of high energy milling and sintering parameters on reactive sintered (Ti, Mo)C–Ni cermets. *J. Alloys Compd.* **2015**, *636*, 381–386. [CrossRef]
18. Li, Y.; Liu, N.; Zhang, X.; Rong, C. Effect of Mo addition on the microstructure and mechanical properties of ultra-fine grade TiC-TiN-WC-Mo_2C-Co cermets. *Int. J. Refract. Metals Hard Mater.* **2008**, *26*, 190–196. [CrossRef]
19. Zhai, W.; Zhu, Z.; Zhou, W.; Nai, S.M.L.; Wei, J. Selective laser melting of dispersed TiC particles strengthened 316L stainless steel. *Compos. Part B Eng.* **2020**, *199*, 108291. [CrossRef]
20. Guan, Y.; Huang, Q.; Peng, Y.; Liu, X.; Kong, Y.; Mao, H. Phase-field simulation of core-rim structure at the early sintering stage in TiC-WC-Ni cermet. *Calphad* **2024**, *84*, 102660. [CrossRef]
21. Xiong, H.; Wu, Y.; Li, Z.; Gan, X.; Zhou, K.; Chai, L. Comparison of Ti(C, N)-based cermets by vacuum and gas-pressure sintering: Microstructure and mechanical properties. *Ceram. Int.* **2018**, *44*, 805–881. [CrossRef]
22. Jin, C.; Plucknett, K.P. Microstructure instability in TiC-316L stainless steel cermets. *Int. J. Refract. Met. Hard Mater.* **2016**, *58*, 74–83. [CrossRef]

Disclaimer/Publisher's Note: The statements, opinions and data contained in all publications are solely those of the individual author(s) and contributor(s) and not of MDPI and/or the editor(s). MDPI and/or the editor(s) disclaim responsibility for any injury to people or property resulting from any ideas, methods, instructions or products referred to in the content.

MDPI AG
Grosspeteranlage 5
4052 Basel
Switzerland
Tel.: +41 61 683 77 34

Coatings Editorial Office
E-mail: coatings@mdpi.com
www.mdpi.com/journal/coatings

Disclaimer/Publisher's Note: The title and front matter of this reprint are at the discretion of the Guest Editors. The publisher is not responsible for their content or any associated concerns. The statements, opinions and data contained in all individual articles are solely those of the individual Editors and contributors and not of MDPI. MDPI disclaims responsibility for any injury to people or property resulting from any ideas, methods, instructions or products referred to in the content.

www.ingramcontent.com/pod-product-compliance
Lightning Source LLC
Chambersburg PA
CBHW062106220926
51850CB00010B/2911
9783725852277